Watson, Lyall.
The dreams of dragons : riddles of natural history / Lyall Watson. -- New York : Morrow, c1987.
178 p. ; 24 cm.
Bibliography: p. 165-170.
ISBN 0-688-06365-9 : $15.45

1. Natural history--Miscellanea--Addresses, essays, lectures. I. Title

THE DREAMS OF DRAGONS

Books by Lyall Watson

The Dreams of Dragons
Heaven's Breath
Lightning Bird
Whales of the World
Lifetide
Gifts of Unknown Things
The Romeo Error
Supernature
Omnivore

THE DREAMS OF DRAGONS

Riddles of Natural History

LYALL WATSON

WILLIAM MORROW AND COMPANY, INC.
NEW YORK

Library of Congress Cataloging-in-Publication Data

Watson, Lyall.
The dreams of dragons.

Bibliography: p.
Includes index.
1. Natural history—Miscellanea—Addresses, essays, lectures. I. Title.
QH81.W438 1987 508 85-32003
ISBN 0-688-06365-9

Printed in the United States of America

First Edition

1 2 3 4 5 6 7 8 9 10

BOOK DESIGN BY JAMES UDELL

Preface

The Kashubians, a Balto-Slavic people still living in parts of Poland, tell of the mysterious Flower of the Fern.

It is, they say, bright red and blooms only at midnight in midsummer, when the earth is "like a child that knows a poem." It lasts no more than a few hours, but it can be picked—with the exercise of due care and a red silk cloth. And it may be carried home, provided that you avoid standing still and refuse to give an answer to any late wanderer who might happen to ask the way. You can be certain that he is not all he seems.

Failure to observe these precautions inevitably results in the disappearance of the flower, and can expose you to the risk of a hideous death at the hands of witches. But success in any attempt to carry the Flower of the Fern home before the first light of dawn guarantees health, happiness, and a long life.

Botanists, in particular, have problems with this story. They are quick to point out that all ferns, by definition, belong to a group of nonflowering plants. A fern flower is a scientific impossibility. And it defies logic to suppose that anyone could know so much about something that nobody has ever seen.

Yet the belief persists—and not just in Poland. Every culture, stubbornly holding to convictions that seem at first glance to be misguided and maladaptive, has its fern flower.

I think such things are worth a second look.

I suspect that many of them reflect a longer lasting, more natural kind of knowing. We ought, instead of simply dismissing unusual beliefs out of hand, to be wondering why they are so widespread and so resilient—and trying to find out what they mean. It is possible that a belief in the incongruous, despite lack of what science would regard as "sufficient evidence," could in itself be a potent social force. In the course of human evolution, a change of mind, a new idea, can have as much survival value and adaptive significance as the mutation of a gene.

In other words, we need our fern flowers, whether or not they exist.

In the meantime, I am besieged by Kashubians. Every mail comes replete with rumors of ferns in flower. But in between the midnight guests, I have time to sow a wild spore or two of my own in scientific soil and see what it is that grows.

These essays are some of the results. Each takes an odd idea, something from the soft edges of science, and tries to nourish it with natural history—to work it, somehow, into the fabric of earth.

You may not approve of the premises, but I hope that you enjoy the products.

LYALL WATSON

Ballydehob, Ireland, 1985

Contents

"Things get odder on this planet, not less so."

—LOREN EISELEY, *The Unexpected Universe*

THE DREAMS OF DRAGONS

1

The Success of Failure

In the beginning, there was an ice-cold current that swept up the west coast of Africa, carrying polar storms and penguins right into the tropics.

It continued for millions of years, unmercifully compelling the climate into shaping a hard, barren, sandy land. For almost 3,000 kilometers from Cape Point to the Cunene River, there is now nothing but rocks and dunes, rugged hills and patches of thorny scrub.

This is the Skeleton Coast, whose history is written in its sun-bleached bones. Walk here, as I did as a child, and you can read every line.

In open graves along dry riverbeds lie the long limbs of ancient reptiles that once scrambled through the Mesozoic, hissing and snapping at the dark. Half-buried above the high-tide line stand cages of ribs that contain the remains of modern whales that found some kind of despair in the deep and tried to hurl themselves back onto the land.

Somewhere between the two rests a phenomenon. A prodigy concealed only by the sand and enormous mounds of broken shell.

I was fifteen when I found it. Just a skinny kid, burned dark by the sun, curious and solitary, very happy on my own. I loved that dry country and knew something of the richness in the wrinkles of its skin. I was familiar with the midden mounds. The fragments of shell shone like beacons, visible from a great distance all along the old raised beaches. But I knew nothing of those who had made them, except that they were sometimes called Strandloopers—Afrikaans for "the beach walkers."

Then, ankle deep in sand one day, I stubbed my toe, dug down, and found a bone. Another tiny fragment in the land, but rather different from the rest. I cleared the area carefully, finding more bones, leaving them in place, and keeping on excavating until I had a hole the size of a large suitcase. And in it, the body of a man.

He was small, smaller than I, with a huge head. He lay on his left side in a fetal position, with his knees drawn up and one hand raised to pillow the great dome of the skull. I had been a little afraid when I first realized that the skeleton was human, but all fear disappeared as I exposed the figure fully and could see how frail and vulnerable it was.

The limbs were weak and spindly and the ribs no thicker than paper. But the head was incredible. Beneath the high arch of the forehead, the face was straight and small with delicate jaws and tiny teeth. It was a child's body, with a childish face, driven by a gigantic brain.

I learned later that his people are known to science as Boskopoid, after the site in South Africa where they were first discovered, and are regarded as interesting if somewhat meaningless freaks, having a cranial capacity 30 percent larger than ours. They are also thought to be ancestral to the Bushmen who still eke out a precarious existence in the Kalahari.

But I have my doubts about both of these assumptions.

The little we know of human evolution seems to show that as our ancestors became bipedal, walking upright and releasing their hands for other functions, they began to carry objects up to their faces for more direct observation. This hand-eye relationship was the beginning of a new, more manipulative kind of intelligence that led to rapid cortical growth. The size of the brain increased and, with it, the size of the cranium. But eventually a point was reached where the head became uncomfortably large for birth by a vertical animal with a poor pelvis. So we brought forward the moment of birth, shortening the gestation period and producing an infant with a head of bearable size, but with an unbearably long period of dependence.

We also somehow managed to slow down the rate of brain and head growth in the last months of pregnancy, and to make up for this deficit by an unparalleled burst of enlargement after birth. In the first year of a human baby's life, its brain triples in size.

With so much effort and energy concentrated up ahead, the rest of the body inevitably suffered and development lagged behind. Childhood was prolonged and some infantile physical characteristics were retained right into maturity. Our teeth still erupt much later than those of our relatives. Our brows are smooth and juvenile instead of ridged and armored, and some of our cranial sutures never close completely.

We have prolonged our adolescence and our lives and changed the whole direction of our evolution toward greater reliance on brain than brawn. It seems inevitable that this process will continue: that there will be a greater development of the brain, a further thinning and extension of the cranium, and an additional attenuation of the jaws and other joints. And given a hundred thousand years, we might end up looking exactly like that gentle being with his huge brain who sat on a remote African beach and watched the ice age coming in.

The man of the future may already have arrived, and lived, and died.

Several of his middens and graves have now been excavated, but it is disappointing work for an archaeologist. There are no traces of any kind of permanent structure, no weapons or tools, no engravings, carvings, paintings, or even any personal adornments. The one concession to technology yet to come to light is a collection of stones suitable for breaking open shellfish too tough for delicate fingers and tiny teeth. The Strandlooper gathered his food on the shore, ate it raw, and slept on the sand. If he had any furniture, it was all in his mind.

Tied as we are to a mechanistic concept of culture, we find it difficult to attribute intelligence to anything or anyone who does not share our kind of configuration. We tend to have more respect for technology than we do for intelligence. Our tests for mind are measures of experience and manipulative ability, nothing more. Those without our hangups, or our hands, will always fail them. The latest research on dolphins and whales, for instance, shows that we have been underestimating them very badly. It seems that they possess an awareness, a consciousness, that may be more than a match for our own. I believe we are making a similar mistake with the Strandloopers.

Like dolphins, they had large brains and yet lacked a material culture. It is almost as though they hadn't any hands. Or didn't need them. Our hands gave us our brains, which in the end seem destined to make hands redundant. So why not short-circuit the whole evolutionary process? Given a big brain, use it. What more do you need?

I have a feeling that things did indeed happen this way. That some genetic frenzy pushed an early branch of human development rapidly through to its natural conclusion. That the Strandloopers got their big brains and were able to keep them intact, out of the cut and thrust of competition elsewhere, because this small group at least existed on a remote uninhabited shore where they could make a relatively easy living.

Here on the edge of the continent, free from the battle for survival going on back in the cradle in the interior, they reached fulfillment. Eating when they needed to and sleeping where they pleased. Playing a lot, because that is something children do. Nimble in their bodies and their minds, because that is the way of adolescence. And they would never grow up. Never become heavy-footed, small-minded, thick-skulled, and morose, because they had chosen a different direction.

I believe this is how it was. I choose to see them playing, sitting, thinking, dreaming; instructing each other in manners and morals; learning grace and loving beauty; finding satisfaction in making ideas rather than machines. I think they may well have discovered trance states and perfected effortless techniques of consciousness raising that we are now only beginning to explore in our laborious and painful way. And I make all these farfetched and unscientific assumptions because there is a secret I share with those walkers of beaches.

When I opened that simple grave in the dunes, there was more there than just the bones of a man. There was evidence of ceremony in the way he had been placed at rest on his left side, the side of the heart and the way of the past. There was a suggestion of future renewal and resurrection in the fetal position and in the way his right foot was half raised as if to begin the dance all over again. There were all these signs of care and concern, but most important of all, for a people who placed so little value on material goods, there was a gift.

Lying beneath the bones of the slim right hand, as though it had been held there in the palm pressed tight against his navel, was a circular white object. I prized it loose and found it was the operculum of a giant snail. The lid that is attached to the side of a mollusk's foot and that it pulls in after the body to seal the entrance to the shell. This particular one was beautiful. On one side it was pure raised pearl, deep and cloudy and translucent, like a crystal ball. The other side was flat, white, and marked with a thin line that wound down from the perimeter in a smooth and perfect spiral to its vanishing point in the center.

It was not an artifact. Such objects occur naturally. It was not manufactured or modified in any way; but it was chosen. The very act of such selection made it special, and the choice of this object for that purpose made it very special indeed. I knew nothing then about votive objects or sympathetic magic, but it felt very sacred to me. I wanted urgently to keep it, but even then I knew enough to put it gently back into the open hand and to fill in the grave and leave it as I had found it.

Much later, I learned about crystals and spirals, about their arcane significance and their use in meditation. Even now, thirty years on, I am only just beginning to appreciate how much there is in simple shapes and forms, and how important they are as keys and cues in the process of universal enlightenment. I still have a lot to learn about even the most elementary levels of consciousness, but as I grow, I cannot rid myself of the certainty that the beach walkers have been there before me. There is no other way to keep a giant brain happy. If it isn't turned outward and involved in experiment and manipulation, then it must be looking in, exploring the maze of consciousness and the marvels of interconnection.

The ice ages were a time of great extinctions. One of them carried these dreamers away; or they died of some poison tide; or fell victim to more aggressive and less thoughtful invaders. Maybe there were never quite enough of them in the first place. We don't know what ended their games on the beach. Certainly they did disappear and must now join a long list of other unsuccessful experiments. Technically they were inferior; empirically they must be regarded as unfitted for survival. Biology requires only one proof of success, and that is continued survival.

But again I have my doubts about this kind of evaluation.

When the blue whale finally succumbs to mindless human predation, it will have become extinct. It will cease to survive. But will it have failed? Can you call a species that has been around for 20 million years a failure? The whales, if they do die, will not die by default, or through some personal failure or deficiency. If they disappear, it will be because we have failed them.

I know that we are earthborn and part of the ecology. That our acts, however thoughtless they may be, are in the natural order of things. In a sense both beach walker and blue whale have been shown to be unfitted for survival. They should have seen us coming.

And yet, if one looks back at the course of evolution, one sees a strange thing. The successes were so busy being successful, looking good, that all of the biggest leaps, the most profound steps, had to be taken by failures.

It was a freshwater crossopterygian, a sort of failed fish, that found competition too intense beneath the surface and in desperation sought refuge on land. We are all his descendants.

Somewhere along the line the mammals began when a timorous

little reptile took to the dark to avoid the rampaging dinosaurs of the day.

Then, as now, there was a future for a failure with flair.

In those days there was room for maneuver, scope for experimentation, and time to start again. That is no longer true. Cockroaches have had 250 million years to get themselves together. Man has had to make do with little more than a million. And by our own endeavors, we have accelerated the process so much that the next step, if there is to be one, might have to be made in just four minutes—the maximum time allowed by our early warning system against nuclear attack.

In a time of flux and indecision, when we desperately need new answers but haven't yet been given the relevant questions, we could use some inspired failures. The beach walkers could have taught us something, but they don't live here anymore. That is sad. More sad than I can say, because I think that something quite irreplaceable died with them.

Happily, however, there are other strange people, other failures with extraordinary gifts still living around us.

They may be hiding in the most remote rain forests of Brazil. Or we may simply have hidden them away from ourselves by locking them up in asylums in Tokyo or Texas.

Whoever they are, and whatever it is that they do, we need them now. Without them it may be impossible for us to recognize that many of the gifts in question are already ours. That we have just been too lazy, too busy, too stupid, or perhaps too much afraid to unwrap them.

So I grieve for the death of the Strandlooper. I mourn the loss of his big brain and I walk a beach or two in search of the vision that moved him. I sense that the answer lies close to home, but I travel a lot anyway. I look and listen in other places, because I learn most about myself from other people. I know that most situations, most environments, are difficult to perceive when they saturate the attention. One needs to look at them, not only from a distance, but also very carefully, because there are pitfalls involved.

I have no difficulty, for instance, in swallowing the saliva in my mouth. I know nothing about it and seldom think of it at all. But there are very few things that could persuade me to swallow it if I had first of all to put it in a glass and examine it carefully as a good scientist should. Detachment and perspective do permit pattern recognition, but they also produce alienation.

I know that an outsider can see things in new ways, producing powerful insights, but I am also well aware of the dangers inherent in the scientific method, which makes outsiders of us all.

Analysis of my saliva in a glass will tell me what it contains, but that won't help me to drink it again. And yet it is the same saliva that I continue to swallow while agonizing over the decision. Nothing has changed and there is not necessarily anything wrong with the sample in the glass, but a wedge has been driven between us. And I am wary of that wedge because it makes an observer of me when I feel that we should now be more concerned with our roles as participants.

We are all involved in the struggle for survival and it is the fate of the Strandloopers that makes me concerned about the destiny of all those other special ones, the exceptions that prove so many rules.

My science, for instance, tells me that the common clover plant belongs to the genus *Trifolium,* which, by definition, has to have three leaves. I appreciate that, and I am grateful to the botanists who have worked out the details of the embryology that produces that triple design, but I owe a far greater debt to a little girl who once took me by the hand to see where a four-leafed clover grew.

My instinct tells me that we need the strange ones—the freak plants and magical exceptions, the child prodigies who begin composing music at the age of four, the idiot savants who cannot read or write but can work out square roots to the fifth decimal place in seconds, the metal benders and the ones who see unusual things.

Instead of getting stuck on surface details, they all respond in some way to the underlying form in things. They are closer to the roots of being, more folded than the rest of us, more touched by continuity. They practice sensory blending; they dance to the light and find inspiration in the earth. We are all a little like that, we all have access to power, but I think we need their example to make it work. We need to know them better.

It really is a matter of maintaining balance. We have lost ours by coming down too heavily on the side of technology and reason, and we have begun to pay the price. But I don't think it is too late to change.

It seems significant that just at this time, when we most need new answers, all kinds of hints and clues are cropping up everywhere. Some are lost in unlikely places and others are ignored or underestimated simply because they have become distorted by some cult or enthusiasm of which we disapprove.

But they are here in abundance right now and, while I would prefer to see us take them seriously, it may not in the end matter how we deal with them. That they exist could be enough.

The Strandlooper, my beach walker with his big brain, exists. He lies right where I left him on that sunset strand. There is no need for me to go back and plow through kilometers of soft sand in a desperate attempt to find him again and set his enormous head up in the scientific spotlight.

I think he has already done his work. The fact that he lived and died is sufficient—and has made us different.

He may not be a direct ancestor of ours, or have passed on any of his genes, but we are the inheritors of his imagination—the dreamers of his dreams. And it is high time we learned how to dream again.

2

The Importance of Pattern

Nobody sees much of unicorns anymore.

Judging by the frequency of their portrayal in medieval art, they were once quite common. But when did you last hear of one in your neighborhood? They seem to have gone to ground or, at the very least, to have undergone a drastic change in behavior.

In an attempt to solve the problem, the American ecologist Lamont Cole turned to a publication that, in the days before computers, was the standard tool for statisticians and those involved in experimental design. He opened a book called *A Million Random Digits,* which used to be put out by the RAND Corporation, and picked an arbitrary place to begin. Then, counting from this point, he selected a series of 8,760 numbers at regular intervals. Let these, he said, represent the level of metabolic activity of a unicorn at the end of each hour in a normal year. There should, of course, be no relationship among the numbers and no cyclic pattern. But what Cole found, to his astonishment, was that unicorns were apparently most active at three o'clock in the morning!

Which accounts for the lack of recent sightings. Unicorns have, it seems (probably as a direct result of persecution by the Royal College of Heralds), become nocturnal. And we are left with some awkward questions about random tables.

A similar dilemma was recently revealed by an engineer working for the General Electric Company. He picked up a book of logarithm tables in the company library and happened, on the way back to his office, to look down at the edge of the closed book. It appeared that colleagues using the book had, for some strange reason, more occasion to calculate with numbers beginning with 1 than with any other number. Which, on the face of it, is absurd. In theory, all nine digits ought to be equally well represented in any calculation. In fact, they are not.

The first digits of random numbers are not equally distributed. Try looking at the numbers in the telephone directory for any large city. Most now have seven-digit numbers in which the first three may be related to a local exchange, and therefore follow a pattern, but the last four ought to be completely random. They never are. The numbers 1, 2, 3, and 4 occur far more often than 5, 6, 7, 8, and 9. The odds on the last sequence beginning with 1, 2, 3, or 4 are in fact seven to three in favor.

So what is going on? A conspiracy involving the telephone company? No. On this count at least, they are innocent. The peculiar distribution, which actually helps to make it a little quicker and easier to use an old-fashioned phone, is in fact an inherent characteristic of the way in which we count. The whole system is biased in the sense that it contains a hidden pattern which means that there is no such thing as pure randomness. It is mathematically impossible to produce a series of numbers that has no pattern at all. The best we can do is to create sequences that have no pattern that anyone is likely to look for—and call these random. Perfect aperiodicity seems to be beyond our grasp. It is so rare, the chances are that it probably doesn't occur in nature at all.

What does happen in most natural situations is that things tend to occur in clusters. Cherries in a fruitcake are randomly, but never evenly, spaced. In theory, there ought to be the same number in each slice. In practice, they gather in groups that most often materialize in the slice that gets served to someone else. Such clusters of allegedly random events are upredictable, but not surprising. That's the way things work.

The chance of someone in a game of bridge being dealt thirteen cards all of the same suit is 1 in 635,013,559,600—but it is still certain to happen. As a combination of cards, it is as likely as any other hand. That 1 chance in 600 billion is going to come up sometime, no matter how surprised you may be that it should have happened in your game. There is no need to write to the newspaper and no justification whatsoever for assuming that "it could never have happened by chance."

It did, and it will happen again. But if the same player on the same evening picked up a second perfect hand with another thirteen cards of a single suit, that would be a different matter. You would have some reason to be surprised or to suspect that someone involved was up to no good. No one would blame you if you left to find yourself another and more honest game. And yet, mathematically speaking, even this raw deal was bound to happen.

Our understanding of the whims of chance is really quite recent and owes a lot to gambling in the last few centuries, although games of chance, rituals for flirting with fate, go back a very long way. Cubes of ivory or stone, their faces marked with symbols or numerals, have been thrown throughout history. The oldest ones were probably astragali, roughly cubical bones from the ankles of goats or sheep, which have been found in sets on many prehistoric sites. They still form a vital part of the prophetic kits used in Africa by priest-diviners.

The trick is to find something that rings the changes, that produces a number of random responses. A cube is perhaps the most simple and satisfactory of all such devices. When falling on a flat surface, it provides an unequivocal comment—only one side faces up. And, if the cube is perfect and fairly thrown, there is an equal likelihood of any of its six faces being uppermost. The outcome is generally recognized as being beyond normal control, but there seems always to have been an awareness of something not entirely random about the results.

The roots of the word *chance* lie in the latin *cadentia,* which means "falling," but has also a sense of rhythm—hence the English word *cadence.* This hints at the existence of pattern, at some kind of beat in the rise and fall of the dice—the presence or absence of luck. At first this would have been interpreted as evidence of the will of the gods. But no game of chance was ever just a game or left entirely to chance. All have ceremonial aspects that suggest attempts to control or influence destiny. Every gambler, ever since the games began, has looked for some way of gaining a personal edge.

Early attempts were largely votive. They concentrated on ways and means of gaining the attention of the gods. But the Renaissance changed all that. The flood of Greek ideas, of classical science and skepticism, into the marketplace of sixteenth-century Italy brought about an entirely new outlook. People began to take control of their own destinies and in this climate it seemed perfectly reasonable for a group of dedicated gamblers to take their particular problems directly to the greatest scientist of their day.

Galileo was fascinated by the behavior of falling bodies. He was in the process of renouncing Aristotle, revolutionizing gunnery, and laying the foundations for the development of modern mechanics. But he was easily diverted by the dice. Their motion, he realized, was subject to the same forces as pendulums and cannonballs and ought therefore to be equally predictable. He joined the game, recorded the patterns of fall and ended up writing a treatise on dice in which, for the very first time, fate was made to bend a knee to science. Galileo tamed luck by laying down the laws. The laws of chance.

The tide of reason flowed out of Italy into France, where something very similar took place in the mid-seventeenth century. A brilliant young mathematician, Blaise Pascal, was commissioned by a gentleman gambler distressed by his losses at dice. Pascal worked out an additional set of rules to help the unfortunate chevalier, checking them as he went along in a lengthy correspondence with fellow prodigy Pierre de Fermat. Between them, they not only brought about an abrupt change in the gambler's fortunes, but developed a system that was refined a century later by a third Frenchman, Pierre Simon Laplace, into the theory of probability.

Their idea was very simple. If any event is certain to happen, they said, let us give it a value of 1. If it is certain *not* to happen, if it is impossible, let its value be 0. All other possibilities can be expressed as fractions between these limits.

For instance, each throw of a dice cube has six possible results. With a perfect cube, none is more likely than another. The chance of any one side—let us say the one with six spots on it—coming to rest face up is one in six. So the likelihood of throwing a 6 first time is one sixth. That is its probability.

The elegance of this approach is that it reduces the vagaries of chance to very simple grammar and arithmetic. The theory deals only with the probability of something happening, regardless of whether it is likely or unlikely. There is no need to talk about improbability. If the probability is less than half, the odds are against that event's taking place.

This new way of looking at things has had an extraordinary effect on the development of science. It absolves us of the need for absolute certainty. It makes it possible to get useful and reliable information from things that are in themselves uncertain. The fall of a single die remains unpredictable, but with the theory of probability, we can predict the outcome of a vast number of falling dice with remarkable accuracy. If the probability of throwing a 6 first time is one sixth, then this also means that in a long series of throws, very nearly one sixth of them will be 6's. Prediction, once the exclusive property of soothsayers, has become a valid tool for scientists.

Chance is no longer quite so chancy. But it is important to understand that this applies only to groups of events. We may be able to measure the probability of a single event's taking place, but we are still as ignorant as ever about its actual outcome. It is the pattern that has become predictable. Sometimes surprisingly so.

The German Army in the last years of the nineteenth century was justly proud of its cavalry. Every regiment kept a large number of horses and, with these in daily use, it was inevitable that accidents would happen. In the twenty years between 1875 and 1894, a total of 196 troopers were kicked to death in fourteen different army corps. This is not at all strange—the surprise is that the pattern of accidents fits perfectly with a formal curve known in probability theory as the Poisson distribution. It was almost preordained.

Murders in England and Wales in this century show an equally astonishing respect for the laws of statistics. In the 1920's, they killed 3.84 people per million in the population. In the thirties, it was 3.27 per million. In the forties, 3.92; in the fifties, 3.3 per million. And even in the easygoing sixties, the "love generation" were killing each other off at the same steady rate of 3.5 people per million of the total population.

According to the statistician Warren Weaver, the situation in the United States is no less orderly. He suggests that "the circumstances which result in a dog biting a person seriously enough so that the matter gets reported to the health authorities, would seem to be complex and unpredictable indeed." And yet, in New York City 75.3 people were bitten every day in 1955. In 1956, the number was 73.6. In 1957 it was 73.2. And in 1957 and 1958, the figures were 74.5 and 72.6.

All of which is more than a little disturbing. How do German horses adjust the frequency of their lethal kicks to conform with an equation? Who stops the Welsh and the English from killing each other beyond a certain level? What connection can there possibly be among the domestic dogs of New York City that prevents them from biting too many people all at once, or sends them out in rabid packs to make up their daily quota?

The behavior of individual horses, dogs, and homicidal humans is entirely unpredictable. But the response of large numbers of each seems to be assured. If disorderly events are sufficiently widespread, they have a strange capacity for becoming orderly. If you can gather together enough uncertainties, the outcome is suddenly certain. Random events, provided only that there are a large number of them, preferably spread out over a period of time, produce patterns with reason and meaning. This is the paradox of probability. It is a mystery that John von Neumann, the inventor of game theory and arguably one of the greatest mathematicians of this century, once described as "nothing less than black magic."

I prefer to see it as part of the pattern that makes life possible. That gives nature its ability to fight back against the laws of thermodynamics and extract order from the general confusion. Life is a rare and unreasonable thing, and has a disturbing habit sometimes of getting things uncannily right.

For instance, choir practice in the small town of Beatrice in Nebraska usually began at 7:20 P.M., but on the evening of March 1, 1950, all fifteen members of the choir were late. The minister's wife, the one who played the organ, was still ironing her daughter's dress. One soprano was finishing her geometry homework; another couldn't start her car. Two of the tenors were listening, each in his own home, to the end of a sports broadcast. The bass had taken a quick nap and overslept. There were ten separate reasons to account for the unusual fact that not one of the choir turned up on time. And at 7:25 P.M. that evening, the church was completely wrecked by a devastating explosion.

If we assume that each of the choir was late for about one in every four rehearsals, then the chance of everyone being late on the same day was about one in a billion. This is improbable, but it is not necessarily surprising. It can happen. That it should have happened on the same night as the boiler blew up is a lot more surprising and begins to border on the uncanny. Coincidences of this order take some explaining.

The Greek physician Hippocrates believed that "sympathetic" elements in the world had a tendency to seek each other out. "There is one common flow," he said, "one common breathing, all things are in sympathy." The Renaissance philosopher Pico della Mirandola also believed that the world was governed by a principle of wholeness. That there was a "unity whereby one creature is united with the others and all parts of the world constitute one world."

The nineteenth-century philosopher Arthur Schopenhauer defined coincidence as "the simultaneous occurrence of causally unrelated events" and went on to suggest that a single event could figure in two or more different chains of circumstance, linking the fates of different individuals in surprising ways. He was the direct spiritual father of the Viennese biologist Paul Kammerer, who in 1919 took the concept a step further with his law of seriality. This says coincidences come in series, clustering together like cherries, not at random, but as a result of some cosmic principle that "takes care of bringing like and like together."

In 1952, the Nobel Prize-winning physicist Wolfgang Pauli and the great Swiss psychologist Carl Gustav Jung collaborated to produce a connecting principle that attributed coincidences to something they called synchronicity, which operates independently of the known laws of classical physics. They suggested that there is a single still-mysterious force at work in the universe—something that tends to impose its own kind of discipline on the chaos required by the second law of thermodynamics, which holds that the natural state of things is a disorderly and completely random distribution of matter.

Jung emphasized the importance of "meaningful coincidences," things that Arthur Koestler was later to call "puns of destiny." These can be as silly as the American journalist Irv Kupcinet opening a drawer in his room at the Savoy Hotel in London to find some personal things belonging to a friend called Harry Hannin—at the precise moment that Hannin, on tour with the Harlem Globetrotters in Paris, discovers in a drawer in his hotel a tie left there several months earlier by Kupcinet. Or they can be as profound as the flow of events that contrived to keep fifteen people alive by making them all late for a date in Nebraska.

Such sequences may be more common than we think.

William Cox, a businessman in North Carolina, used a simple and elegant method to test the idea that some people might be able to avoid traveling on trains that were going to be involved in an accident. He collected statistics on the total number of passengers who had traveled on the same train during each of the preceding seven days and on the fourteenth, twenty-first, and twenty-eighth days before the accident. His results, which cover several years of operation with the same equipment at the same station in a variety of weather, show that people did indeed seem to avoid accident-bound trains. There were always fewer passengers in the damaged and derailed coaches than would have been expected for the train at that time. It seems that some people, consciously or unconsciously, are sufficiently uneasy about a situation of that sort to find reason not to travel. Or that Jung's principle is strong enough on its own to interfere with the plans of those people in ways that make them miss the train.

The American parapsychologist Rex Stanford tells of a New Yorker traveling by subway to visit friends in another part of the city. Instead of changing trains at the Fourteenth Street station, he absentmindedly walks out into the street and bumps directly into the people he was going to see. Stanford suggests that behavior of this kind, which brings about the right result with the wrong moves, is determined by access to information at an unconscious level, producing what he calls psi-mediated instrumental responses. Which is an awkward name for an important idea. It is possible that when the need arises, we can scan our environment for information that would not be available to our normal senses. Such an ability would certainly have survival value and will have been actively selected for and encouraged during the course of evolution. Why then is it not more common? Why isn't everyone using it? How come there were any people at all on the train by the time it had its accident?

The answer has to be that "psi," however it works, is not a genetically determined characteristic like blue eyes or curly hair. It is not subject to the usual selection pressures, it doesn't obey the rules and get passed on to successive and grateful generations, because it is not encoded in our chromosomes. It is available to us, and probably to other species, some of the time and under certain circumstances, simply because we and they are part of a process that has patterns of its own. And when you look at these closely, it becomes apparent that the whole universe is extraordinarily finely tuned.

Most of what goes on in the physical world is determined by a small number of universal constants, of things that do not vary. The value of these constants is now well known, but it is only recently that we have begun to realize how many of them are the result of wildly improbable coincidences. The same arrangements keep cropping up again and again in different natural contexts, as if the supply of suitable patterns was extremely limited. The mathematics need not concern us here, but it is a fact that the pattern that mathematicians describe by the expression $\Delta^2 V$ arises in connection with gravity, light, sound, heat, magnetism, electrostatics, electric currents, electromagnetic radiation, waves at sea, the flight of airplanes, the vibration of elastic bodies, and the mechanics of the atom. The applications may differ, but the pattern is identical and is a source of delight to mathematicians who, in the words of the creative French adept Jules Henri Poincaré, "practice the art of giving the same name to different things."

The most widespread objects in the universe are neutrinos, stable chargeless particles that pass right through earth at the speed of light. Their mass has been calculated as 5×10^{-35} kilograms, which is extremely small, but needs to be very precise. If it had been 5×10^{-34}, this fractional increase would have been enough to bring the expansion of the universe to a halt long before now.

The structure of stars depends on their gravity. Large ones radiate heat at a great rate, turning themselves into "blue giants." Small ones are cooler, relying more on convection currents, and are known as "red dwarfs." Both are unstable, but between these two extremes is a very narrow range of star sizes that permit stability of the kind seen in our sun, whose luminosity has changed very little over 4 billion years—allowing us, among other things, the luxury of a liquid ocean. If gravity were very slightly weaker, all stars would be red dwarfs. A correspondingly tiny change in the other direction, and all stars would be blue giants. And in either case, there would be no planets and no life as we know it, anywhere.

These examples of cosmic cooperation (and there are many more) are compounded by a series of numerical coincidences. The number of stars in a typical galaxy is the same as the number of galaxies in the universe. And there is an astonishing similarity between the age of the universe and the number of particles in it. Both hover, along with a long list of other basic parameters, around the huge and mystic figure of 10 to the 40th power (10^{40}) which led the Nobel Prize-winning English physicist Paul Dirac to conclude that "something strange is going on."

Dirac and the astronomer Sir Arthur Eddington were prompted by these delicate balancing acts to make connections between them and our own existence as conscious creatures. They and others have pointed out that even a slight tilt in the turning would have made our kind of life and our existence impossible. John Wheeler, professor of physics at Princeton, goes further by suggesting that there may even be a causal connection. "Here we are," he says, "so what must the universe be?" The implication of this is that the very fact of our existence constrains the structure of the universe, selecting what it must be like. At first sight, this sounds ridiculously egocentric, or more precisely geocentric, taking us back to a pre-Copernican situation in which the earth was seen as the center of the universe. But what is meant is that life is strange, something that requires very special circumstances, and that these circumstances might be atypical in the sense that they arise only locally and for a comparatively short space of time, when the patterns of the universe happen to coincide with our needs.

The fact that we find ourselves living on solid surface, when the vast majority of the material in the universe takes the form of gas clouds or hot plasma, and that we happen to be near a stable star, when many others are unstable or unable to support planets in any form, is not coincidental. It couldn't have happened in any other way. In other words, we *are,* and (for this moment in time and at this place in space) things have to be the way *they* are. There *is* a connection.

We are very lucky to be here. Whichever way you look at it, the chance of all the necessary conditions coming together in just the right way is very small indeed. Unless you include the notion of design.

A hole in one at golf is, on the face of it, highly unlikely. There are vastly more locations on a golf course than the hole in the middle of the green and, in theory, a ball driven at random is equally likely to land on any of them. All are equally improbable. But the fact is that in golf, that hole in the green has a very special significance that is not possessed by other random patches of grass—which is why everybody claps when the ball arrives in the hole. It has what amounts almost to a magnetic attraction, gathering in the ball eventually, even if it takes more than one stroke to get it there. The ball, in effect, has designs on the hole.

Life, perhaps, is part of such a pattern. It had designs on the earth. And happened here because it intended, or was intended, to do so. Probability theory allows life, however improbable it may be, to evolve. In fact, it requires it to do so eventually—just as it demands the existence of naturalists who must write erratic essays about the extraordinary way in which nature has conspired to arrange its affairs for the benefit of living things.

I can't help feeling, though, that there is more to it than that. I have some trouble believing in a Great Golfer, but I feel a deep sense of satisfaction at the way things are. I am touched by the magic. The patterns please me and I find myself, more and more often these days, wide awake at three in the morning, standing at my window on the off chance of a glimpse of a great white horse with an improbable horn.

3

The Roots of Awareness

I hadn't thought about it much. Not until the day in 1968 when I climbed the stairs of an old office building within sight of New York's tawdry Times Square.

I was on my way to see Cleve Backster, ex-intelligence agent and expert on the use of lie-detecting equipment. He had retired from the CIA and now made a living teaching police officers how to use the official polygraph. But old habits die hard and instead of squeezing the truth out of criminal suspects, Backster spent most of his time interrogating various vegetables.

This strange preoccupation had begun on a February morning two years before our meeting, when Backster found himself alone with his polygraph and time to spare between classes. The machine is a relatively simple one designed to do little more than register small changes in the electrical conductivity of a subject's skin. The theory is that these electrodermal responses are sensitive indicators of emotional change, and that a skilled operator can detect minute differences in the patterns traced by a pen recorder on a moving chart and decide which of these are caused by the stress of telling a deliberate lie.

Backster was very good at his job and wondered whether he was skilled enough to be able to monitor the changes that ought to take place in the electrical conductivity of a plant as additional water rose from the roots of the leaves after a shower of rain. He brought in the office plant, a fine healthy *Dracaena massangeana,* and attached the polygraph electrodes to either side of one of its broad green leaves. After balancing the leaf into his circuitry, Backster watered the potted plant. Nothing happened. So, in frustration, he decided to try what he calls "the threat-to-well-being principle, a well-established method of triggering emotionality in humans." In other words, he decided to torture the plant.

As a first step, he dipped the tip of one of its leaves into a cup of hot coffee. There was still no reaction, no flicker on the recording tape, so he decided instead to get a match and burn the leaf properly.

"At the very instant of this decision, there was a dramatic change in the tracing pattern in the form of an abrupt and prolonged upward sweep of the recording pen. I had not moved, or touched the plant, so the timing of the response suggested to me that the tracing might have been triggered by the mere thought of the harm I intended to inflict."

Backster left the room to find some matches, and discovered that his return caused another jump of the recording pen. But when he actually held the flame up to the leaf, the reaction was much less marked; and when, later, he merely pretended to set about burning the leaf again, there was absolutely no reaction at all.

To Backster, there seemed only one possible explanation. The plant was reading his mind. It seemed to be able to differentiate between his real intentions and pretended ones.

The next step was a logical one for a law enforcement officer. Backster brought in a partner, his assistant Bob Henson, to help play the old "good-guy, bad-guy" routine. They picked on an inexperienced subject, another common house plant, in this case a young *Philodendron cordatum* from the office next door, and Henson did his best to terrify the potted plant. And every so often, Backster would come in, put a stop to the harassment and speak to the plant in soothing tones.

It seemed to work. By the time I arrived to interview Backster, the philodendron appeared to be able to distinguish between the two men. Every time Henson, the "bad guy," entered the room, there was an agitated response on the polygraph, which disappeared as soon as "good guy" Backster came near—or even spoke on the telephone in an adjoining room. Then the recorder slid easily into a gentle flicker that even I could recognize as being very similar to the response of a human being undergoing a pleasant, relaxing experience.

Cleve Backster first talked about this discovery on a New York radio interview in 1967 and amongst those who heard him was Pierre Paul Sauvin, an electronics engineer with the International Telephone and Telegraph Corporation in New Jersey. Sauvin designed a silicon-chip potentiometer that was a hundred times more sensitive than Backster's polygraph and, working something like the dimmer switch attached to an electric light, was able to register far finer shades of response. Through this, Sauvin hooked up his own plants to the big green eye of a cathode ray oscilloscope and a bank of sophisticated recorders.

Sauvin discovered that he got the strongest responses from plants that he had himself raised from seed, or with which he felt a special rapport. And that the more time he spent with these plants, feeding them, touching them, or washing their leaves, the more dramatic their reactions became. He attached a loudspeaker to an answering machine that made it possible for him to talk to the plants by telephone at regular intervals during the day. And soon found that everything that happened to him, even something as trivial as the electrostatic shocks he received from his metal desktop on dry winter days, was mirrored by a measurable response on the equipment attached to his plants 5 kilometers away. A weekend holiday with a girlfriend produced recordings of several unprecedented peaks that, as far as he could remember, corresponded exactly with every shared orgasm in that lakeside cottage over 100 kilometers from the plants at his home.

Meanwhile, back at Times Square, Cleve Backster had extended his experiments to over twenty other species of plants. Like Sauvin, he found that the plants seemed to be aware of his personal crises, responding strongly on one occasion when he applied iodine to a cut finger. He became convinced that he was dealing with some sort of sensitivity that operated at a very fundamental, perhaps even a cellular or molecular, level. He called it primary perception and set up a classic experiment to prove its existence.

From an aquarium shop, Backster bought a supply of brine shrimp (*Artemia salina*)—little crustaceans that are widely sold for feeding to tropical fish. He decided to use the death of these shrimp as the experimental signal and, to preclude any possibility of human influence, built a mechanical device that would, at random intervals, dump one or more live shrimp into a pot of boiling water. As test subjects, he chose three philodendron plants brought in from outside sources that had never been tested before, handled them as little as possible, and kept them in separate rooms under conditions of constant light and temperature.

The experiment worked. The plants responded to the death of the shrimp five times more often than could be explained by chance alone, and Backster concluded that "there exists an as yet undefined primary perception in plant life that can be shown to function independently of human involvement." He published the results in 1968 in *The International Journal of Parapsychology.*

The response was staggering. Seven thousand scientists and students asked for reprints of the paper, features and stories appeared in dozen of newspapers and magazines, and gardeners everywhere began to feel less self-conscious about being overheard talking to their plants.

I was fascinated by the reports. As a biologist, I was intrigued by the evolutionary significance of the discovery. What did it mean in terms of survival? If dying shrimps do indeed send out a distress signal, why on earth should it be of any interest to a potted plant?

I was aware of alarm signals among social animals. Sea gulls, for instance, have specific calls that warn their breeding colonies of the approach of predators. Ground squirrels and prairie dogs have an early warning system that alerts their societies to the danger of air raids by birds of prey. The function of these signals is so clear that those of birds such as crows have been recorded and broadcast across airfields to frighten them off the runways just before planes were due to land. Very often the alarm is interspecific. Terns, starlings, and pigeons feeding with the gulls all take flight at the sound of a gull alarm call; and seals dive into the water when nearby colonies of cormorants give notice of approaching danger.

Alarm calls obviously have high survival value and work well across the species line, but not all species function on the same frequencies or even with the same sense organs, so there could well be a strong natural pressure toward the evolution of a common signal—a sort of all-species SOS, warning perhaps of a common danger such as the approach of an earthquake. Needs of this kind in evolution seldom go unnoticed and it seemed to me that Backster's discovery could be nature's answer to precisely such a pressure.

I agreed with Backster that a signal accessible to all life would have to be very basic. Something that every living thing could receive and understand. All organisms consist of cells and therefore a search for the source and reception of such a signal ought to be made at a cellular level.

On my first visit to his office, Backster demonstrated this possibility very vividly by scraping a few living cells from the inside of his cheek and killing them in a glass dish by the addition of a drop of dilute sulfuric acid. At the moment of their death, Backster's favorite philodendron reacted with what, in a human subject, would have been described as mild alarm.

I repeated this test later with a potted plant and a blood sample of my own and found that I got a good response, but that the results were even more satisfying when I worked with a specimen of semen. Live sperm are perhaps the only part of a human body specifically designed to lead an independent existence outside the body, and might therefore be more likely to be put in a signal situation than cells from the inside of the cheek. Sperm seem to be more easily and more strongly alarmed, altogether more responsive than mundane red blood cells. And, quite apart from anything else, they are a lot more fun to collect.

Later still, I went on to experiment with other people, with some equally fascinating results. On a lecture tour of colleges in the

United States, I played a variation of the old parlor game called Murder, in which six people are chosen at random and draw secret lots that designate just one of them as the murderer. In my botanical version, this person is required, when no one else is watching, to kill one of a pair of identical potted plants. The surviving plant, the sole "witness" to the murder, is then taken into protective custody and attached to a lie detector. Each of the six suspects is then brought back to the scene of the crime individually and confronted with this witness for the prosecution. What often happens is that the polygraph/plant combination produces no significant response to five of the six, but "accuses" the other one of the crime—and is very often right.

On only one occasion, at a small country college in Florida, the machinery seemed to regard two of the six suspects as equally guilty. On later cross-examination, it turned out that one was indeed the culprit. The other had spent three hours, before coming in to classes that day, felling a stand of trees on his family farm. He came in, as it were, with blood on his hands.

It is very tempting to conclude not only that plants are aware of human emotions, but that they have memories and perhaps even some sort of emotional life of their own. There is a wealth of folklore and anecdote to support such an assumption. Everyone knows someone who has a "green thumb," who seems to be able to get plants to grow well just by being around, by talking or even playing music to them. In Colorado, petunias and marigolds grew abundantly and flowered earlier than usual when serenaded by Bach's chorale preludes, while an identical planting in a second climate-controlled greenhouse nearby wilted and died when exposed to a barrage of constant hard-rock music. In Canada, a farmer growing corn and soybeans was able to increase his yield of both crops almost 20 percent by broadcasting musical selections over his fields. The soy, it is said, seemed to be particularly partial to Gershwin.

It is difficult to know what to make of these results. Much of the work, including Backster's and my own, is poorly controlled and open to criticism on the grounds that the electrical responses being observed could be artifacts inherent in the equipment or produced by the environment. A study at the University of Washington has found that even very small fluctuations in temperature or humidity can produce changes in plants, altering their electrical resistance in ways that will register on a polygraph. And several attempts to replicate the critical experiment with brine shrimp—most notably one very carefully controlled series at Cornell University in 1975—have failed to find any significant response at all.

But it is equally difficult, for those of us who have taken part in a broad range of experiments with plants, to dismiss all the results quite so easily. The responses observed on recording tapes or oscilloscopes may not be the result of any direct communication between humans and plants. The fact that measurable changes occur just as experimental stimuli are being applied could be coincidental. But it happens so often, and at such appropriate moments, that it is hard to avoid the conclusion that some kind of direct response is taking place. It seldom happens when there is no plant involved and never occurs at all when both plant and human are removed.

The fact that some people get results where others fail with the same equipment or procedure may be significant. Dr. Ken Hashimoto, managing director of Fuji Electronic Industries in Kamakura, Japan, tried to repeat Backster's work without success until he let his wife, who is renowned for her green thumb, use his apparatus. When Mrs. Hashimoto stands anywhere near the cactus that her husband has connected to his equipment with acupuncture needles, there are strong reactions, and when she talks to the plant, it seems almost to make measured and appropriate responses.

The response of orthodox botanists is quite unequivocal. They dismiss the whole subject as pseudoscientific nonsense. But things have been happening in the field of plant physiology during the last few years that make the idea of plant sensitivity and memory much less fantastic.

Plants, it seems, are much more like animals than the textbooks would have us believe. They have no nervous systems, but they do make use of electricity to transmit signals, and they behave at times almost as if they were acutely sensitive.

Charles Darwin was one of the first to consider the matter in detail. He noticed that garden plants changed the position of their leaves at night so that lupines, for example, hung miserably down, while radish leaves moved into a vertical position soon after dark. He suggested, in his book *The Power of Movement in Plants,* that this change from the normal horizontal position prevented dew and frost from settling on the leaves and saved them from the danger of nocturnal chilling. His ideas fell into disfavor when it was pointed out that the same "sleep movements" take place in plants growing in areas where frosts do not occur. But in 1982, a botanist at the Scripps Institution in California proved that horizontal leaves are always cooler at night than vertical ones and that even this small difference in temperature inhibits growth in the horizontal ones. He showed that bean seedlings grow far better on a good night's sleep.

Darwin also did painstaking work on insectivorous plants. His favorite was *Dionaea muscipula,* the venus flytrap, which he called "the most wonderful plant in the world." He noticed that the two sides of the trap would snap shut like the pages of a book only when one or more of a group of fine hairs in the hinge were stimulated in a certain way, and assumed that some sort of signal must pass rapidly from these "triggers" to the cells that produced the actual movement. Darwin showed the plant to a medical physiologist working on the electrical properties of animal muscle, and together they were able to record distinct electrical impulses sweeping through the leaves of the plant when the trigger hairs were touched.

These signals travel at a speed of about 2 centimeters a second, which is very slow compared to the 100 meters a second typical in higher animals, which have special nerve cords capable of rapid communication. But even these relatively slow speeds enable some plants to respond sufficiently quickly to catch a fly or to ensure fertilization.

In the common garden shrubs *Mahonia* and *Berberis,* the male organs lie tight against the cup-shaped petals of the flowers, but at the slightest touch these elastic stamens spring out to slap pollen onto a passing insect. The Australian trigger plant *Stylidium* has its male and female parts fused into a fingerlike projection that hangs enticingly from the flower, and when this is touched it whips through 180° in less than a hundredth of a second. This is the fastest known movement made by any plant and it succeeds in both covering an unwary visitor with pollen and fertilizing itself with pollen picked up by that insect during an earlier ambush elsewhere.

Perhaps the touchiest of all vegetables is the sensitive plant *Mimosa pudica.* Any contact with the rows of leaflets sends them folding up like fans. A hard knock against the stem makes whole leaves droop, and an overt injury can result in the complete collapse of an entire mimosa bush. This behavior has been familiar since it was first recorded by Chinese scholars over two thousand years ago, but it is only in this decade that we have begun to understand it. A group led by Barbara Pickard at Washington University has shown how the movement begins with an action potential—a short-lived negatively charged electrical wave that signals to the rest of the plant that all is not well.

In some less advanced plants, this serves merely to isolate an injured area while repairs are underway. In others, such as the tomato, the whole plant responds by closing its stomata—the little windows through which a plant breathes—helping in this way to conserve moisture until the wound is healed. But in a select few species, the whole plant takes dramatic avoiding action. *Mimosa pudica* has become so sensitive that it responds to changes in temperature, direction and intensity of light, electric shock, cuts, burns, and even changes in atmospheric pressure. It is rumored that a mimosa at Kew Gardens in London was touched so often by curious visitors that it had a nervous breakdown and shed all of its leaves!

The survival value of such behavior seems clear. It might not deter a determined goat, but a sudden movement is quite likely to disturb a potential insect pest. At the University College of North Wales, botanists have been working on *Biophytum sensitivum,* one of the sorrel family, whose movements are not as dramatic as the mimosa's, but which exhibits an even more interesting sensitivity. They have discovered that while the plant responds to a light touch by letting its leaflets droop, it is also capable of responding even before it has been touched. All that is necessary to produce the avoidance reaction is to hold a charged Perspex rod 2 centimeters above the leaf surface.

It is known that flying insects become highly charged in flight as a result of their rapid wing beats. And it appears that *Biophytum* has learned to react to the presence of such an electrostatic field in the air by taking appropriate avoiding action even before an approaching insect has a chance to land. If it can do this, then the possibility of reaction to, and interaction with, the fields produced by some humans no longer seems quite so outrageous. Perhaps those of us who seem to have green thumbs are simply better grounded than others, giving off none of the negative action potentials that stimulate some plants into taking evasive action and remind others of traumas produced by old wounds and injuries.

There is real evidence now to show that plants actually do have such memories, and use them. At the University of Clermont in France, bur marigolds (*Bidens pilosus*) are being raised with a memory of childhood injuries. Researchers there wait until the plants produce their first two symmetrical seed-leaves and then deliberately wound one of these cotyledons by giving it four deep pricks with a needle. Five minutes after this treatment, both cotyledons are removed completely to ensure that the result involves memory and not a response to an ongoing wound.

The marigolds bud normally and are allowed to grow until they are twenty days old, and then measured. In untreated, control plants there is an equal chance of either of the two usual cotyledon buds developing into the main growing stem. But the treated plants seem to "remember" the asymmetrical pricks received and respond by growing strongly in the opposite direction—away from the side that was injured.

This is a clear example of fixed long-term memory of the sort common to many animals. But the marigolds also seem to have a labile short-term memory of the sort that enables us to remember a telephone number while dialing it, but to forget it immediately afterward when involved in some new concern. When the first four pricks on one cotyledon are followed within fifteen minutes by a single prick on both seed-leaves, the plants seem to forget the earlier indignity and concentrate on the later injuries, showing no tendency to grow more strongly toward one side or the other.

News of this work, which was published in a Danish botanical journal in 1982, is only just beginning to spread through the scientific establishment. So far it has had surprisingly little effect. It will be a while, perhaps not until someone else has repeated the work elsewhere, before psychologists appreciate that what we have here is a pattern of behavior that ought to be amenable to classical conditioning. If plants have memories, then they can presumably be trained to respond to special stimuli—they can be shown to be capable of learning.

Given that plants are living organisms, that they grow, respire, and respond to changes in both internal and external conditions, it is strange that we still have difficulty in accepting them as behavioral organisms. We seem to suffer from a vegetable prejudice that makes it hard for us to attribute to them characteristics that we readily allow to other animals. But, as understanding grows, it is getting harder and harder to deny plants equal rights to some sort of sentience. It cannot be too long before there are institutes of plant psychology!

In 1983, botanists in New Hampshire began to look at the way in which poplar and sugar maple trees respond to attacks by leaf-eating caterpillars. Both species react by producing protective chemicals such as tannins, which make them indigestible and less attractive to insect predators. Seedlings of the two trees were grown under controlled conditions in the laboratory and some were deliberately damaged, having their leaves cut and torn to simulate insect attack. As expected, these plants doubled their chemical defenses within two days, but what the scientists had not been expecting was that the same changes took place in other seedlings that were left untouched.

To explore this situation further, the experimenters put plants into separate pots, so that there was no root contact among them, and arranged them in ways that prevented all direct contact among their leaves. But still the undamaged plants continued to mimic the defensive reactions of those that were being deliberately mutilated.

Now a second team of botanists, working in the woods near Seattle, have discovered that a similar situation exists among willows and alders. The natural defense of these plants to attacks by tent caterpillars and webworm is also to produce chemicals, alkaloids, that make them unpalatable. They do this in direct response to actual attack, but, like the poplars and maples on the other side of the continent, trees that have not yet been infested show evidence of taking preemptive precautions as well.

Both groups of scientists conclude that, in addition to the distasteful chemicals, the damaged plants must be releasing pheromones, airborne hormones that are carried on the wind like alarm calls, warning trees in other parts of the forest of the danger of imminent attack by insect predators, and leading them to make their own preparations to meet the onslaught.

The only thing preventing us from classifying such trees as social creatures is the fact that the ones receiving the warnings are incapable of running away. But it becomes difficult to deny that, given their relative lack of mobility, the actual response of the plants is highly appropriate and very impressive. There is a level of environmental awareness in evidence here that makes it hard to go back to looking at plants as stolid, unresponsive systems. In fact, the only way that many plants can be made to conform to the narrow view of them is to actually knock them out. And this can be done in precisely the same way as one would go about anesthetizing a human or an animal.

A whiff of ether has precisely the same effect on a sensitive mimosa. Under the influence of the gas, the plant is oblivious to even the most violent attempt to get its leaves to cringe. But as the anesthetic wears off, the mimosa slowly recovers and begins again to respond, first to strong stimuli and finally, as usual, to even the most delicate touch.

Chloroform has similar effects on water plants, which become comatose and stop producing their usual stream of oxygen bubbles. Barbiturates slow down all normal processes, preventing sleep movements and stunting the growth of rice seedlings—which seem to become as addicted to them as human insomniacs. And gardeners everywhere have long known that aspirin has a vitalizing effect on plants, fighting infection, stimulating leaf growth, helping conserve moisture, and keeping cut flowers fresh for far longer in a vase.

As far as we know, there is no botanical equivalent of a headache. Plants lack a true nervous system or a brain. But their ability to store and carry memories, and their capacity for meaningful communication, suggest that the mechanics of information transmission and retrieval may be far more fundamental than many scientists now believe.

It begins to seem possible that many of the talents and abilities we have been reserving for "higher" animals may in fact be part of the experience of all living things. And that, in our attempts to understand the origin and nature of awareness, we ought to be looking far more closely, and more literally, to our roots.

4

The Nature of Crowds

In 1813, the American ornithologist John James Audubon watched a flock of passenger pigeons (*Ectopistes migratorius*) so large that "the light of noonday was obscured as if by an eclipse." The birds passed directly over his head and kept on coming at the rate of 300 million an hour for several days. He estimated that the entire flock consisted of more than 20 billion birds stretching for over 1,000 kilometers across the plains of Wisconsin.

A century later, there wasn't a single passenger pigeon alive anywhere. Indiscriminate shooting and trapping wiped them out and the last one died in the Cincinnati Zoo in 1914.

One of the greatest wildlife spectacles of all time has gone, but it is still possible to get some idea of what it must have been like by visiting certain African water holes on the edge of the Kalahari Desert shortly after the start of the rainy season. At this time, a little sparrowlike bird called the red-billed quelea (*Quelea quelea*) congregates to breed. The colonies are so dense that nests hang five thousand to a tree over an area of more than 10 square kilometers. The eggs in most of the nests hatch at the same time and when they do, the broken shells come tumbling down like snowflakes.

Within a few weeks, the whole quelea colony—adults and young—takes to the air. Millions of birds spread out across the land, but soon they bunch into local flocks tight enough to block out the light. Then these dense brown-feathered clouds, whirring like helicopters, descend to feed or drink, banking and wheeling with extraordinary precision, acting like gigantic organisms and reacting together with speeds that defy description.

The mechanism of communication between the individual birds remains largely mysterious, but there are obvious reasons for getting together in such crowds.

Birds that join a flock benefit immediately by being given greater security against predators. They enjoy the freedom to feed in relative peace without having to keep looking over their shoulders for signs of danger. With thousands of pairs of eyes in a flock, hawks and foxes stand little chance of getting close without being observed. So, though each individual can afford to be less vigilant, the flock as a whole is more responsive to sources of threat. And though the size of a flock might make it look unwieldy, the fact is that a crowd reacts more quickly to threat than any individual can.

Zoologists at the University of Rhode Island have measured the reaction time of common starlings (*Sturnus vulgaris*). Each bird in their study was placed in a chamber designed to have no confusing echoes and exposed to sudden sharp bursts of pure sound at random intervals. An electronic apparatus recorded the bird's response and showed that, on average, an isolated starling reacts with minute muscle twitches in just 80 milliseconds—that is, 0.08 second. But in a flock of starlings in flight, maneuvers seem to take place almost three times as fast. Frame-by-frame analysis of film of such flocks shows that waves of reaction shoot through the group, passing from bird to bird in just 30 milliseconds—0.03 second.

This apparently supernatural ability has been a source of amazement to biologists for many years.

In 1931, the English naturalist Edmund Selous watched jackdaws (*Corvus monedula*) rising simultaneously from fields where they were feeding and suggested that the only way to explain such instant coordination was by assuming that something he called thought transference was taking place. Reluctant to accept the possibility of telepathy in birds, others have tried since then to show that perfectly ordinary signals involving movement or sound were being sent and received. But it is not easy to do so. There is so much general noise going on in a large flock of birds that specific acoustic cues such as warning cries seem unlikely to be the whole answer. And the birds at the back of a big crowd are so completely hidden from those at the front that it becomes equally unlikely that they operate on the simple visual basis of signals such as head movements or wingflaps that call for everybody to follow the leader.

In 1974, the Rhode Island team came up with an ingenious suggestion that some species at least might respond to electromagnetic radiation. They point out that migrating birds are known to be sensitive to earth's magnetic field, using it to help them find their way home—and that a receiver capable of picking up signals as weak as these should be able to detect stronger and closer currents created by movement through the air of other birds in a flock. We know that in among their flight feathers, many birds have specialized hairlike structures called filoplumes, which are rich in free nerve endings. These could act as short whip antennae for picking up such signals, but it is difficult to prove that they do.

If true, this intriguing idea may also help account for the speed with which schools of fish respond to outside stimuli. Electromagnetic signals would certainly be carried through a group even more quickly underwater—especially seawater, which is an excellent conductor. Now, however, there is news of another study that offers an equally interesting theory—and this one has the merit of being simpler and far more easily tested.

On the coasts of Europe and North America, large flocks of little wading birds called dunlin (*Calidris alpina*) feed along the shoreline and, when startled, take to the air in a dense mass of hundreds of individuals that behave as though they were of one mind. They look for all the world like a single wheeling airborne creature that has the startling ability to change color instantly, turning from light to dark and back again as the birds pivot in unison, showing first their dark backs and then their white bellies. This rapid alternation as the flock turns from side to side often occurs in less time than it takes a solitary bird to react to a simple stimulus.

Wayne Potts of the University of Washington filmed dunlin in action around Puget Sound in 1984 and discovered that aerial maneuvers always begin with a single bird on the edge of a flock turning into the mass of its fellows. The bird nearest this instigator reacts comparatively slowly, taking 62 milliseconds to follow suit. But the next one gets the message more quickly, and the one after that faster still, so that by the time such a "maneuver wave" reaches the far side of a flock, the reaction between birds is four times as fast, taking place in just 15 milliseconds.

This sounds amazing, but in fact there is a human equivalent. Chorus lines of high-stepping stage or ice dancers do something very similar. Human reaction time is, on average, around 200 milliseconds—0.2 second. But films of chorus lines in action show that rehearsed maneuvers—even ones begun without warning—travel down the line at almost twice the speed, jumping near the end of the row from one person to another in just 100 milliseconds—0.1 second. The explanation for such "supernatural" sensitivity is that the individuals on the end of a row of performers can see the wave of action coming and are able to anticipate its arrival, making their kick or spin well ahead of normal time.

It looks as though birds and chorus girls have more in common than their customary feathers. And that for both there is not only safety in numbers, but access also to aesthetic and sensory displays beyond the reach of any individual.

There are purely mechanical advantages as well. Studies on the aerodynamic interaction of aircraft flying in formation have shown that they have pronounced effects on each other. This is because any flying object uses two kinds of power. The first is called profile power—which is the direct energy needed to overcome *drag* and drive an object through the air. And the second is induced power—which is a more indirect source of energy provided by *lift,* a kind of buoyancy given to objects once they have started to travel through the air. Profile power is constant for a plane or bird that is flying at its normal cruising speed, but induced power depends on the stability of the air around it—and varies enormously. Birds and planes flying in formation can give such power to, or take it away from, each other.

Theoretical studies made at the California Institute of Technology show that there is an upwash beyond a wing tip that can be useful to another wing—as long as the second system is close enough to take advantage of it. In other words, two birds flying close abreast can give a little help to each other. And if three birds fly in a line, the one in the center can be even better off, getting a boost from either side. According to the Cal Tech formula, the benefits go on growing with each additional bird added to the line, reaching a peak with about twenty individuals flying along side by side on a common front. In this formation, a flock can travel 70 percent farther than a single bird—without any of the individual birds in the flock using any more energy.

And in practice, this is precisely what happens. Pelicans, ducks, geese, and cormorants on long journeys always fly in close-knit groups of around twenty individuals. But they seldom fly in line abreast or in a single column one behind the other. Why? Well, the Cal Tech theorem shows that with more than ten individuals in line-abreast flight, the center birds enjoy twice as much uplift as the ones on the ends of the line. Which is hardly fair. But if the line is bent into a vee formation, the birds on the ends of the two arms receive additional upwash from all those in front of them, which nicely cancels out the disadvantage of their position. Waterfowl, in fact, always take on vee-shaped flight formations, adjusting for whatever inequities may still exist by changing places in the line from time to time.

Ornithologists have long wondered if the bird at the head of such a formation was necessarily the strongest and most experienced, but the Cal Tech study shows that this position at the apex of the vee is no more strenuous than any other—and is in fact often given to a weaker or younger bird that needs a little extra uplift en route.

Fish become social, gathering together in crowds, for somewhat similar reasons. As long as their air bladders are well adjusted, giving them almost neutral buoyancy, they don't have to worry about staying afloat—but there are definite dynamic advantages to being in a school.

As it travels through the water, a fish leaves vortices—little whirlpools—behind it on either side. The ones to its left travel clockwise and the ones to its right counterclockwise—which means that the flow within them, on the edge closest to the fish, is opposite to the direction of travel. So any other fish getting too close to a school member will find that it is in an adverse current. But if it moves just far enough away to swim through the far side of a fellow's vortex, it will find itself in a favorable current. Which is, of course, exactly what schooling fish do. They position themselves precisely one vortex-width apart.

None of this of course means that birds or fish are able to understand abstruse mathematics or the principles of dynamics. They simply fit in. If a goose moves ahead of or falls behind the vee line in flight, it finds that it has to use more energy just to keep up. If a fish gets too close to or too far from its schoolmates, it gets left behind or eaten. But by taking the line of least resistance, by letting themselves be shaped by natural forces, these creatures inevitably arrive at appropriate and meaningful adaptations. They make good environmental sense.

There is a natural principle in action in such situations that has never been spelled out in biology. We appreciate the obvious advantages of social behavior, but we haven't yet come near to any real understanding of what happens to individual plants or animals once they become part of larger societies. In almost every case, the whole is very much larger than the sum of the parts and behaves in ways that can seldom be predicted from a knowledge of, or an analysis of, nothing but the components.

This getting-together-for-the-greater-good has been going on for a long time, producing some extraordinary results.

There is, for example, a creature without head, tail, limbs, mouth, or body cavity that nevertheless flourishes in most parts of the world ocean. It is one of a group of jellyfish that look like distinct individuals, but are actually colonies of very simple animals, both adults and larvae, which cling together to make a complex whole. The combination is a blob of blue gel consisting of buds that serve various functions. Some catch food, some digest it, some are concerned with sensing light and dark, some are sex organs, and a few take the form of thin trailing tentacles that can be as much as 50 meters long.

This weird assortment of animals somehow work together. They share no nervous system, but electrical signals are conducted through the mixed community to produce astonishing concerts of rhythmic and apparently harmonious activity. The societies behave. They have the sort of coordination and unity of purpose we normally associate with individuals. They do things together, working toward a common end—a little like the collection of individual instrumentalists that go to make up a well-integrated symphony orchestra.

The conductor in this case is trying to construct a transparent sky-blue air-filled float that turns the aimless colony into a creature that travels with a purpose. With the float inflated like a sail, the blob of gel becomes a Portuguese man-of-war (*Physalia physalis*) that skims across the surface of all the tropical waters of the world.

When there is no wind, the man-of-war's sails are deflated. But when the wind blows, the float is pumped up to take advantage of the breeze and the colony starts to work as a well-trained crew, trimming the sail by fitting its curvature precisely to the wind. Then the man-of-war sets sail. And the most wonderful thing of all is that it does not just go wherever the wind happens to blow. It sets its own course. The blob navigates!

The long tentacles stream out behind like a sea anchor, while the rest of the colony arrange themselves in a clump on one side of the float so that the beast as a whole is asymmetric. The man-of-war floats with its long axis, including the sail, at an angle of 45° to the wind. In nautical terms, it sails "on a broad reach."

In a light breeze, a man-of-war travels at about 4 knots—120 meters a minute. Which means that it can cover 10 kilometers or more in a day—as long as it doesn't become entangled in floating weed. In fact, it seldom does, because weed tends to collect in lines at right angles to the direction of ocean currents—which flow at fixed angles to the prevailing winds. In the Northern Hemisphere, the currents and the weeds are deflected to the right of the wind. But the men-of-war employ a different tactic. The majority of the sailing colonies north of the equator bulge to the right of their floats, which means that they sail at an angle to the left—directly between the windrows of weed. And in the Southern Hemisphere, where the wind and currents are reversed, most of the colonies are right-sailers and enjoy a similar freedom.

I am impressed when things like this happen. Or, to put it more objectively, when natural selection results in relationships between living things and their environment that make such wonderful sense. Evolutionary theory requires that it be this way, that the fittest, most sensible creatures survive at the expense of those that are not as well adapted. But sometimes the fit is so precise, and the procedure used to achieve that fit so roundabout and unlikely, that I can't help wondering what else might be involved.

It is not necessary to look to the ocean or the deserts for further examples. There are equally bizarre relationships in our own history and behavior.

The Nobel Prize–winning novelist Elias Canetti once wrote a fascinating book called *Crowds and Power.* It was an inspired work that ought to have had a profound effect on social psychology, but it was published twenty-five years ago—long before the current debate over sociobiology—and seems to have been largely ignored. Which is a pity, because it has a lot to teach us about ourselves.

Canetti regards the crowd as an organism in its own right. He makes a basic distinction between crowds and random groups of people—such as those waiting for a train—who just happen to be at the same place at the same time. A genuine crowd is a group of people who gather round a focal point, which he calls a crowd crystal.

Crowds can occur anywhere, and often get together almost without warning. At one moment a street is empty save for a random scattering of individuals. And in the next, in response to a mysterious signal that travels between people like electrical news through the flesh of a jellyfish, there is concerted action. People begin to rush together to form the nucleus of a crowd. Those involved in the action seldom know what has happened. If stopped and questioned, they are unable to provide any reasonable reply, but they nevertheless hurry toward an invisible but common goal, wanting only to be where most other people seem to be.

Canetti describes a crowd in this preliminary phase as "a nebulous entity feeding on people." It is, he says, in a juvenile stage and driven by only one instinct—the urge to grow. It wants to be bigger. It needs more people and seizes all those within reach, absorbing them indiscriminately into its body. Anything shaped like a human being can join and, once engulfed, becomes an equal part of its substance, part of the crowd.

A crowd creature exists for as long as it has an unattained goal. This may be a short-term one—such as a strike, a killing, or the destruction of a building. Or it may be as long term as the vision of the Promised Land that kept, and still keeps, the Jews of the world—the Children of Israel—together.

Species of crowds can be classified according to the nature of their goals, but all have certain attributes in common. Within the confines of the crowd, individual people lose their separate identities, their names and social status, becoming equal parts of the new being. This feeling is so strong that it is possible, in cultures right across the globe, that all demands for justice and all theories of equality are based on the actual experience of brotherhood familiar to anyone who has ever been part of a crowd.

In the density of a crowd, all concepts of individual space are abandoned and all fear of touch and contact is lost in a rare togetherness. In a dancing crowd at a festival or Mardi Gras, individuals become fused into single creatures—strange beasts with fifty heads and a hundred arms and legs all thrashing about in unison. In long-lived crowds with long-term goals, the behavior of the organism may be diluted and controlled. It becomes domesticated by ritual and tradition, channeled toward some national or religious goal, keeping itself alive and going for centuries at a time by creating boundaries that separate Us (in the crowd) from Them (outside it). A closed crowd sacrifices its chance of growth, but gains in staying power, reassembling only as often as is necessary to remind its members of their connections.

Eternal crowds of this kind have changed human history, but in natural-history terms the most interesting and energetic crowds are still the spontaneous ones that rush together for instant gratification. These species reach their goals quickly and die. A strike that manages to achieve any gains crumbles visibly, losing part of its reason for being. A lynch mob melts away, almost apologetically, as soon as the job is done. But before it does, at the moment of success that provides its goal and signals its end, such a crowd often produces a distinctive sound. When the appointed executioner holds up the severed head of the victim—which can be as symbolic as a signed agreement or a desired resignation—the voice of the crowd is heard. This unique cry, the call of the organism, expresses its unity more powerfully than anything else. It is perhaps the most vivid demonstration of the fact that the community of a human crowd is something qualitatively different from the simple sum of its individual parts.

Studies of the mechanism of such cohesion are still in their infancy. They owe only so much to the sort of links apparent in flocks of dunlin or lines of chorus girls. We haven't yet begun to describe what really goes on in human crowds or to look at their natural history and behavior in any detail. At this stage it seems to be beyond us to even predict where and when crowds are likely to occur—except perhaps at soccer matches or other public events with emotional overtones where the ingredients of a crowd are gathered into a dangerous circle where they can see each other and egg each other on. And we have as yet no rules to help us deal with unruly crowds when they do form and run riot.

Disturbances caused by crowds are becoming more and more common as population densities increase and closed crowds begin to erupt, sometimes with great violence, into the open. There are signs that the long-term crowd has freed itself from some of the constraints of religion and tradition and is no longer likely to be content with pious promises. It wants to experience for itself the strongest possible expression of its own animal force and, to this end, seems prepared to accept the flimsiest of social pretexts. And if such outlets are denied, crowds can explode in meaningless rage or disintegrate in a lethal panic.

We need to understand the nature of the beast a lot better. And it seems to me that the best way of doing this is to go back to first principles—to look at how crowds first began.

We need to look again at the way in which all living cells seem to be derived from colonies of formerly independent creatures. We need to understand how some protozoans and mollusks still persuade free-living algae to take up residence inside their bodies and go on living there, turning the hosts green, making food from the sun. We must understand how separate cells in the colony of some jellyfish bodies coordinate their behavior by means of concerts of electricity. We will have at some stage to rearrange our vocabulary to accommodate the reality of hives of bees or nests of ants, deciding whether to refer to such integrated colonies as "them" or "it." We are going to have to decide whether our own bodies are in fact composed of colonies of cells in exactly the same way—and whether we have any right to be considered as independent organisms. And only when all this has been done will we be in any position to begin to understand what happens when two or more of us get together in a crowd.

I suspect that we have more in common with passenger pigeons than is at first apparent; that human crowds and societies have distinct personalities of their own; that together we assume complex new identities and acquire some surprising skills; and that, unless we are terribly careful, these powers could get out of hand and lead us, like those unhappy pigeons, to rapid and untimely extinction.

5

The Source of Concern

In the winter of 1940, a twelve-year-old boy lay in the Myers Memorial Hospital in West Virginia recovering from an operation. It was a stormy night and Hugh Perkins was watching the snow pile up outside the window, when he was startled by something more substantial beating against the pane. He called the nurse on duty and when she opened the window, a pigeon fluttered in.

It was not just any pigeon, but a racing bird Hugh knew well. Several months earlier it had arrived exhausted in his backyard and, on being fed and cared for, had stayed and become a pet. But that was in the summer and in Summersville, 112 kilometers away from the hospital in Philippi. "Look at his leg," the boy told the nurse; "it has a ring on it with the number one sixty-seven." She did, and it had, and the hospital allowed him to keep the bird in a box beside his bed.

When the Perkins family came to visit Hugh a few days later, they confirmed that it was indeed his pet and had been seen around the house for several days after he was admitted to hospital. So it hadn't come cross-country with him, or simply followed the family car. It seems, from the evidence, that the pigeon actively sought out the boy and succeeded, somehow, in traveling over 100 kilometers to do so. Not only that, but it found the correct window, in the right building, in a strange town, at night and in a snowstorm.

The newspapers are full of such anecdotes.

There is a delightful account in the *Irish Times* of equally strange behavior by a pack of foxhounds. No sooner had the dogs been released near Kildare in January of 1959 than they "stormed out of covert like raging furies. The volume of their full cry was so remarkable that it was obviously one of the best scenting days of the whole season. They set such a pace that sixty followers had to ride like equal fury to keep in the same parish with them." The pack charged cross-country, through a wood, up a hill, and over the main Dublin-to-Kilcullen highway, covering 10 kilometers in just half an hour. Then they leaped over a 2-meter-high wall surrounding the graveyard of Carnalway Church—and stopped.

"That was as far as they went. The great cry that they had kept up unremittingly for the previous half hour was now silent. The terrific pace they had maintained over seven miles of big country had now slowed them to a walk. The scent that had been breast-high had vanished completely. One of the most expert huntsmen in Ireland tried every artifice known to the experienced practitioner, but to no avail. A remarkable hunt was over."

It ended on the grave of Major Michael Beaumont, late master of the Kildare hounds, who had died and been buried there just three days earlier. He had often remarked that his dearest wish was that his beloved dogs should, at some time or other, hunt near where he lay. And so, it seems, they did. On their very first outing since his death, the entire pack howled their way at full speed straight from the kennels to their master's grave, and then fell silent.

There is the story of the greyhound Cesar, who followed his master from Switzerland to the court of Henry III in Paris in the sixteenth century. Of the dog named Prince who managed to find his way from England across the Channel to his master's side in France during the First World War. Of Sugar, a cream-colored Persian cat with a distinctive hip deformity, who traveled across 2,400 kilometers of rugged land from Anderson in California to his owner's new home in Gage, Oklahoma. And of King, a Belgian sheepdog who made a 1,400-kilometer journey in the opposite direction, from Sandpoint in Idaho to Richmond in California, right across the Nevada desert, arriving with bleeding paws.

Andecdotes abound, many of them involving animals with marks or injuries that made recognition more reliable. It seems clear that, on occasion, pet animals will go to extraordinary lengths to be reunited with their human associates, often at considerable risk to their own lives. How they do so is a mystery. Another is why they should bother at all. What is it that sends a pigeon out into the night in a snowstorm, or drives a cat for months on end across a desert, several canyons, and the width of the Rocky Mountains?

It is not enough to argue that instinct is the answer. It is true that there are good examples of animals using navigational aids with little or no apparent training. Pigeons find their way with the help of familiar landmarks, subsonic sound patterns associated with the passage of wind through certain mountains, the height and position of the sun in relation to an internal clock, the patterns of the stars, and the lines of earth's magnetic field. We know too that certain species, such as the European cuckoo (*Cuculus canorus*), are able to make these voyages without instruction, following parents they have never known thousands of kilometers south to the traditional winter feeding grounds in Africa. But none of these talents would seem to be able to account for the ability or the need that sent a cat like Sugar on a thirteen-month journey across the highest mountain range in the United States.

Pet owners have no difficulty with the problem at all. It is quite clear to them that the cat or dog involved was motivated by love for its owner. They may be right. The human emotions of loyalty and affection may well turn out to be appropriate stimuli for other birds and mammals, perhaps even for the occasional crocodile, now that these reptiles are known to show elaborate concern for their broods. But emotions are explanations that are unavailable for investigation. It is impossible to provide scientific proof for the existence of love even between two human beings. Another problem with the love theory is that it still provides no mechanism and is totally confounded by cases of reverse movements—like that of the cat called Mastic, who trekked 250 kilometers back to his former house, now occupied by new and unfamiliar tenants, when his original owners moved.

All of which leaves us with Cesar, Prince, and Pigeon 167 in an uncomfortable sort of scientific limbo.

J. B. Rhine at Duke University in North Carolina was one of the first to take the problem seriously. He argued that it is dangerous and shortsighted to deny that unusual behavior patterns exist, merely because we cannot identify the physical mechanism that sets them in motion. In 1962, he and Sara Feather gathered together a collection of fifty-four cases—twenty-eight dogs, twenty-two cats, and four birds—in which "an animal, separated from a person or mate to whom it has become attracted, follows the departed companion into wholly unfamiliar territory and does so at a time and under conditions that would allow the use of no conceivable sensory trail." They called this behavior "psi-trailing."

One of the birds on their case list was a pigeon raised with its mate in a cage kept at a cottage on Redondo Beach in California. The family moved 800 kilometers away to San Francisco, taking the pigeons, who now had a nest full of newly hatched chicks, along with them. Halfway to their new destination, the male pigeon took off and was seen heading back toward Redondo. The family waited some time for him to return, and, when he did not, continued sadly without him. They arrived in San Francisco after dark and hung the cage at one of the windows of their new home. At dawn the next day, to everyone's delight, the male pigeon was found sitting outside the window and was promptly reunited with his brood.

It is important to distinguish behavior of this kind from ordinary homing. Many animals possess remarkable abilities for finding their way back to well-known haunts. A shearwater (*Puffinus puffinus*) that was removed from its nest burrow in Britain and taken 4,500 kilometers away to the east coast of the United States returned in just thirteen days. California newts (*Taricha torosa*), little amphibians that breed in streams on the slopes of the Sierra Nevada, were taken from their pools and released 5 kilometers away in a deep and unfamiliar canyon. The following year, almost all of them had returned to their home pool, climbing in the process over an intervening rocky ridge more than 300 meters high—a feat equivalent, in human terms, to scaling a mountain the height of Everest.

The ability to orient in this way seems to be largely inborn. Monarch butterflies (*Danaus plexippus*) migrate each winter from Canada down through the United States to the highlands of Mexico and return along the same routes the following spring. The boldly patterned hordes fly over 100 kilometers a day, stopping each night to rest in huge brown clusters among the branches of particular trees—the same trees every year, which people along the way have come to believe must possess special spiritual properties. And the monarchs make this annual migration, and unfailingly pick out the traditional motel trees, despite the fact that the butterflies traveling in any year are first- or even second-generation offspring of those that covered the course in the opposite direction the previous season.

All these patterns may be determined by genetic programs—the new all-purpose explanation for strange behavior. But there is no known gene, no established physical force, no electromagnetic link that can act in this way to guide a cat or pigeon to a totally unfamiliar place. New homes and strange hospitals and graveyards are selected by the humans involved for purely artificial reasons and have no biological, geographical, or traditional features that could be a help to natural history.

Psi-trailing seems to be evidence for the existence in other species of a faculty sometimes described in our own as extrasensory perception or ESP. If this is true, it is too important to ignore. The Rhine case histories are all spontaneous ones, involving special situations that can be verified only up to a point. They depend, in the end, on the veracity of those involved and cannot be repeated. There are, however, some obvious and repeatable tests just crying out to be made.

The usual pigeon-homing experiments involve taking a pigeon away from its loft, but it would be just as easy, and very much more revealing, to take the loft away from the pigeon. Obvious sensory clues would have to be carefully controlled. Smell, for instance, could be eliminated by moving the loft downwind. Separating birds from their families and friends in this way would provide a strong and basic biological need to be reunited. A need that could call subtle powers into play. If the pigeons in such a test consistently succeeded in finding their mobile home, and did so in ways that could be shown to be independent of chance, we might be a little closer to understanding how it is possible for the concern of one individual for another to be projected across space between them in ways that appear to be paranormal.

In 1972, scientists working at the Novosibirsk Medical Institute in the Soviet Union announced the discovery of what they called distant intercellular interactions between tissue cultures. They took two samples from the same tissue culture and sealed these securely in separate glass containers. A poison or lethal virus was included in one of the containers and the two were placed in "optical" contact, separated only by the quartz glass walls of their jars. The contaminated culture naturally sickened and died. But twelve hours later, so did the control sample, displaying precisely the same symptoms, despite the fact that the design of the experiment seems to have precluded the possibility of normal infection between the samples.

The implication is that some sort of "sympathy" exists between similar cells, despite their physical separation. Which is an outrageous idea. Or at least it would be, if there were not other evidence to suggest that something of the kind might well take place.

Crystals are fascinating systems—regular geometric forms of matter that seem to arise spontaneously and then go on to replicate themselves in a stable manner. Some substances crystallize easily. Some can be persuaded to do so with a little more difficulty. And a few, it seems, may never form crystals at all unless something untoward happens. Glycerin, for example, was discovered 250 years ago as an extract of natural fats. It was obtained as a sweet, colorless, oily liquid and immediately put to good use in medicine, lubrication, and the manufacture of explosives. But nobody could get it to crystallize. Despite supercooling, reheating, and all the other usual forms of persuasion, glycerin remained resolutely liquid and it was assumed that the substance had no solid form.

Then, in the early years of this century, something strange happened. A shipment of glycerin in transit between the factory in Vienna and one of their regular clients in London went through a severe storm in the Bay of Biscay. When the barrels were opened in London, one of them had crystallized. The client was upset, for it was no good to him in that condition, but chemists were delighted. They began to borrow bits from the barrel to seed their own samples and soon glycerin crystals were being produced, without difficulty, in a number of laboratories. In fact, it happened almost too easily. Soon after receiving a sample of the new crystals in the post, one American laboratory found that all of their stock of glycerin began to crystallize spontaneously—despite the fact that some of it was still sealed in airtight containers.

The conventional explanation is that microscopic, invisible, airborne seeds of the new crystals must have infected all the samples. And when crystallization of glycerin began to take place in distant laboratories, it was claimed that visiting scientists must have contaminated the experiments there by bringing in the guilty seeds in their beards. But the appearance of crystals in sealed containers suggests the action of some other kind of influence. It begins to look almost as though, in addition to the new physical seeds, the crystallization of glycerin is facilitated also by a sort of mind seed, by the knowledge that it can be done—and that this change of attitude has effects that are independent of distance.

Rupert Sheldrake, an imaginative plant physiologist from Cambridge University, has produced a theory to account for such peculiarities. It is possible, he suggests, that the form and pattern of things depends not so much on physical laws as on habit. A substance that has never before been crystallized, for instance, can in theory take on a number of equally stable crystal forms. But once it has, for whatever reason, chosen a form, it will go on producing those particular crystal patterns forever after. In other words, it is easier to do something that has already been done.

Sheldrake calls his theory formative causation and suggests that such things fall under the control of "morphogenetic fields," which are independent of space and time. A crystal in a laboratory in Brazil grows in a certain way, not necessarily because it has been seeded, but because it resonates in sympathy with another such crystal that exists, or once existed, in London or anywhere else. The developing embryo inside an egg takes on the form of a chicken, partly because it is programmed to do so by its DNA, but also because it has the example of existing chickens elsewhere to follow.

"Imagine," says Sheldrake, "an intelligent and curious person being shown a television set for the first time. He might at first suppose that the set actually contained little people, whose images he saw on the screen. But when he looked inside and found only wires, condensers, transistors, etc., he might adopt the more sophisticated hypothesis that the images somehow arose from complicated interactions amongst the components of the set." This, suggests Sheldrake, is how conventional biology still looks at life.

The hypothesis of formative causation, on the other hand, recognizes that the images in the TV set depend on invisible influence from a distant transmitter. The theory appreciates the importance of the components in the set, or of the DNA and other proteins in a living creature, but it also recognizes the role of influences from outside, from other organisms of the same species that are transmitting information on the appropriate wavelength.

This resonance between things is an important idea. It begins to make some sense of otherwise apparently coincidental or supernatural connections.

The difficulty of ordinary communication with submarines has long obsessed naval engineers. Seawater is a very effective shield against almost all radio waves at any frequency, and scientists in several countries have been battling to find some new way round the dilemma. Stimulated by their success with "intercellular interaction," the Soviets turned to rabbits instead of radio. They took newly born rabbits down in a submarine and kept the mother ashore in a laboratory with electrodes implanted deep in her brain. At intervals, the underwater rabbits were killed off one by one, and at the precise time that each of her offspring died, there were sharp electrical responses in the brain of the mother. It is only in recent years, with the use of ELF (extremely low frequency) antennae many miles long, that the world's great naval powers have found a way to keep contact with their distant submarines, but rabbits seem always to have had some sort of natural resonance that keeps them in touch in times of crisis.

Folklore is filled with human equivalents. Ian Stevenson at the Virginia School of Medicine has recorded the case of Mrs. Hurth, a local lady who allowed her five-year-old daughter to go off on her own to meet the rest of the family at the neighborhood movie theater. The mother remained in her kitchen at home, until "quite suddenly, while I held a plate in my hand, an awesome feeling came over me. I knew that Joicey had been hit by a car. I dropped the plate and prayed aloud, 'Oh God, don't let her be killed.' " Mrs. Hurth called the theater, where an astonished and distraught ticket seller confirmed that the girl had in fact just been knocked down, but wasn't seriously hurt and was calling for her mother.

Following up the Russian rabbit experiment, Sergei Speransky in Novosibirsk tried separating mice from a tightly knit social group and discovered that some kind of connection seemed nevertheless to persist. He starved the isolated mice and found that the rest of the group, as if in sympathy, ate significantly more than an intact control group. Then he killed the exiles and found that the survivors in the original, but depleted, group began to breed more rapidly, producing more offspring more often than they had when their group members, though separated, were still alive.

The results of these suggestive, but somewhat brutal, tests have now been confirmed by a far more elegant and humane experiment designed at the University of Utrecht. In 1972, Sybo Schouten trained several Dutch mice to press one of two levers in their cage when the lamp above that lever was lit. The reward was a drop of water, which could not be obtained in any other way. Pressing either lever at other times produced nothing. When the mice were accustomed to this regime, he put one of them into a cage containing lamps but no levers. A second mouse was placed in a cage with levers but no lamps, in an isolated room in another part of the building. The lamps were switched on at random by a binary selector and neither mouse received any water unless the proper lever in one cage was pressed at the precise time that the relevant lamp in the other cage was lit. Success, and survival for both mice, depended on the mouse with the lamp being able, in some way, to communicate with the lever-pressing mouse at the appropriate moment, giving them both the necessary reward. The mice succeeded in getting enough to drink, coordinating their signals and responses in ways that got the lever pressed when its matching lamp was lit, about two hundred times more often then could have been expected from chance alone.

There is evidence to suggest that empathy of this kind may be involved in the acquisition and spread of some behavior patterns among wild animals. In 1952, dairies in Great Britain began to deliver milk in bottles with metal foil caps. These were less sturdy than the old cardboard ones and blue tits (*Parus caeruleus*) in London soon discovered that they could peck through the foil to get at the rich cream floating on the milk—and before long the doorsteps of London and most of the adjacent counties were littered with foil and milky splash marks. The new habit spread steadily through the south of England, extended west and north, and then suddenly seemed to explode. By 1955, all of the blue tits and most of the great tits (*Parus major*) in Europe were doing it. Once enough birds had learned the trick, once a certain critical threshold had been crossed, the habit seemed to jump great geographical gaps, starting up spontaneously in new areas by what Sheldrake calls morphic resonance.

He calls attention to a series of experiments done in the 1920's at Harvard University by the celebrated psychologist William McDougall. These involved putting rats into a tank of water that had two gangways. One was brightly lit and led nowhere; the other was dimly lit and led to freedom. The first rats to be tested took a long time, some of them over three hundred soakings, before they learned which way to go. But McDougall kept on experimenting for fifteen years with a further thirty-two generations of the same rats—and found, as he predicted, that each generation learned more quickly. The average number of errors in the first generation was over two hundred, but in the last ones this was reduced to just twenty. McDougall concluded that the rats had inherited their improved ability directly from their parents.

This work was repeated years later in Australia, where they measured the rates of learning in fifty successive generations of the same breed of rats. These tests confirmed that later generations learned more quickly, but they also made two surprising discoveries. They found that all their rats were smarter than McDougall's to begin with, and that the rate of improvement was the same whether they used the offspring of trained rats or took their subjects from a population that had never been given the test at all.

Sheldrake suggests that this is morphic resonance at work. "If rats are taught a new trick in Manchester, then rats of the same breed all over the world should show a tendency to learn the same trick more rapidly, even in the absence of any known type of physical connection or communication. The greater the number of rats that learn it, the easier it should become for their successors." He believes that this is why children today seem to have higher IQ scores and are learning younger and faster than their grandparents did. And that we have access to what is effectively a cumulative memory.

In a neat test of his theory, Sheldrake has been looking at nursery rhymes. He reasons that it should be easier to memorize sequences of words, even in a foreign language, if these have already been memorized before by millions of people. The Japanese poet Shuntaro Tanikawa provided him with three short Japanese rhymes with a similar sound structure. One of these was meaningless—it was just a jumble of unconnected words. Another was a newly composed verse, and the third was a traditional rhyme known to generations of Japanese children. Sheldrake wasn't told which was which, but soon found that all the English-speaking subjects who were tested were able to memorize one of the three rhymes far more easily than the other two. It was, of course, the popular nursery rhyme.

It is a long way from cultural patterns of this kind, which unite human beings across continents and through time, to an explanation for the links between humans and other animals that make psi-trailing possible. But there could be a connection.

In 1952, Karlis Osis worked with cats in the Rhine laboratory at Duke University. He put his cats, one at a time, into a T-shaped maze and asked human agents to try to will them to turn either left or right, according to a random sequence. The humans sat in an isolation booth behind a one-way mirror so that they could not be heard, seen, or smelled by the cats. Some of the cats behaved purely randomly, as though unaware of any influence, but a few cats produced exceptionally high scores, apparently responding directly to the wishes of the human agents. And in every single case, the most successful combinations were between cats and humans who had already got to know each other well and had an existing bond of affection. There was "resonance" between them.

Sheldrake is convinced that this resonance is nonmaterial. It does not depend on any known kind of energy, and is therefore independent of distance. The evidence we have for apparent telepathic contact between people suggests that this is true. It certainly most often involves people who are genetically similar, as in the case of identical twins; or between people who have developed strong emotional ties, like lovers, or mothers and their children. These kinds of people resonate most strongly with each other. There is abundant anecdotal, and a certain amount of controlled laboratory, evidence to show that such awareness is truly shared, most dramatically in crisis situations. But there is, as far as I know, no suggestion that this sensitivity is directional.

An individual may be aware that someone he knows well is in trouble, unhappy, or thinking of him—he may even have a surprisingly clear image in his mind of the environment of the other person—but no one seems to be able to point in a particular compass direction and say where the information is coming from. Which is what a stranded dog or cat or pigeon must do in order to find an erstwhile owner.

To navigate successfully, you need three things—a clock, a compass, and a map. There is abundant evidence to show that living things possess internal clocks, ticking away in time to natural rhythms. There is equally good evidence to show that many species, ourselves included, have enough iron concentrated in parts of our bodies (in humans it seems to be the bones of the nasal area) to enable us to respond to earth's magnetic field. And we know from experiments with homing pigeons that they, and a number of other species, are capable of learning to recognize landmarks, building up quite complex maps in their memories. But there is something else you need if you are to find your way to a novel destination. You have to know a name or be given the proper coordinates—you need a mark on the map.

I have no difficulty in accepting the thesis that a pet dog could develop enough rapport with its owner to resonate in sympathy with him even at a distance—and to be moved to restore more direct and personal contact. I see no problem with the notion that such a dog could, as we know birds and bees do, pinpoint its own position in space with the help of celestial cues and follow a compass course. There is on record the case of a collie called Bobbie who was abandoned in Indiana and traveled 3,000 kilometers home to Oregon. A reporter later retraced this epic six-month journey, finding people who had given food or shelter to the dog along the way. He discovered that it picked a straight and reasonable route with very few detours, taking what was essentially the shortest way home. But I still have difficulty with the notion that such an animal could apply these skills, as Sugar the Persian cat did over 2,400 kilometers of rough travel, while heading toward a totally unknown and apparently unpredictable destination.

We are left with a persistent mystery. There has to be something else involved. Some factor that even Rupert Sheldrake hasn't had the nerve to imagine. I am comforted by the knowledge that, at some level, I broadcast to those I love, letting them know that I am alive and well. But I am slightly outraged, I find my essential privacy violated, by the thought that somewhere deep inside me is a treasonous little beacon, a homing device, that chatters away on unconscious frequencies, shouting, "Over here! This way! No, no—a little more to the left. Yes, that's it. Right behind this rock . . ."

6

The Dreams of Dragons

On a grassy slope high above the Flores Sea in Indonesia stands a simple white-painted wooden cross.

Fixed to it is a neatly printed plastic plate, the size of an index card, which reads:

IN MEMORY OF
Baron Rudolf von Reding Biberegg
born in Switzerland the 8 August 1895 and
disappeared on this island the 18 July 1974
"He loved nature throughout his life"

And he proved his love by an act that, for a conservationist, has to be the closest thing to dying in a holy war. He let himself be eaten by an endangered species.

The island in question is called Komodo and he was, to be specific about it, devoured by a dragon.

It is nice to know that it can still happen.

Paleontologists insist that the last of the dinosaurs died out more than 60 million years ago—which is at least 50 million years before anything like a man walked the earth. And yet dragons, and stories of encounters with dragons, feature prominently in a variety of cultures.

The arch that King Nebuchadnezzar built in Babylon in the sixth century B.C. was decorated with glazed brick bas-reliefs of *sirrushu*—unmistakable dragons with horns, scales, and talons, which seem to have been drawn from life and to have figured prominently in Babylonian art since 2800 B.C.

Dragons with huge teeth and claws are among the most ancient symbols of Chinese folklore, used to represent both earth and water—often together in the form of fertilizing rain. Every watercourse and well has its dragon, subject to the dragon kings of the four seas that surround earth. The Mountain of Dragons near modern Beijing stands in the center of the old earth and it was here, by coincidence, in 1929 that the skull of Peking man—the first human to be associated with fire—was found.

The Dragon is one of the four great constellations in Chinese astronomy, in which the New Year is celebrated on the appearance of the moon before the rising of the Dragon Star. In the traditions of Thailand, Burma, and Japan, dragons are beneficent animals that ascend to heaven, where the pressure of their feet on the clouds causes rain. Kings and emperors are often said to have the Dragon's Face, given control over storms and the power to take on any shape they please.

Beginning with Greek mythology, the Spirit of Storms is embodied in monstrous chimeras, which take on a variety of forms, but consistently have claws and a scaly tail and breathe fire. And it is this, more terrifying manifestation that seems to move through most of Western thought.

The dragon is the Devil in Christian symbolism. It is discovered in the shrine of Mars by Saint Philip, revealed in Antioch to Saint Margaret, fought in the desert by Saint Michael, and finally conquered in England by Saint George. But dragons still abound in Britain.

They figure prominently in church art, carved in wood and stone, cast in bronze and silver, shown worshiping Christ, being trampled underfoot by him, or simply blazing defiantly over doors and from stained-glass windows. Throughout the Middle Ages, local dragons kept popping up in ponds and woods, eating dogs and persecuting people. The last of these troublemakers was killed as recently as 1668 in Essex and given the ultimate accolade in British eyes—that of a pub sign in its honor.

All these accounts of dragons are so vivid, and the depictions and descriptions so consistent, that it is tempting to look behind the myth for some foundation in fact. For real and recent evidence of dragons in action. And the best evidence, as I have already suggested, seems to come from Indonesia as recently as 1974.

East of Bali is a chain of islands that lie scattered like steppingstones between the continents of Asia and Australia. These Lesser Sundas are separated from Bali by nothing more than a channel 25 kilometers wide, but this is nevertheless a formidable barrier. The water is over 1,000 meters deep and lashed by tides that race through from the Pacific to the Indian Ocean at a terrifying 15 knots.

A few wild boars and sambar deer, both of which swim strongly and well, have managed to make the crossing and establish breeding groups on some of the eastern islands, where they have nothing to fear from tigers, leopards, jackals, or predatory cats. No large carnivores at all have made it across, but the pigs and deer don't have everything their own way. Their move into the Lesser Sundas has been a little like time travel, taking them back to an age when there were no other placental mammals, and marsupials held sway. But it has also taken them back to an age of ruling reptiles—back into the jaws of dragons.

We do not know what killed off the last of the true dinosaurs. Theories range from chronic constipation to showers of radiation released by the explosion of a nearby star. But we do know that, ever since they died, the earth has been taken over by smaller, faster, more adaptable, warm-blooded mammals. Everywhere, that is, except Australasia (Australia, New Guinea, and the islands on their continental shelf), which has been isolated so long that it has no modern mammals. It is decorated instead with wombats, numbats, bandicoots, and wallaroos—living fossils that continue to exist simply because they are protected from competition with later, more aggressive invaders.

Most of these marsupials feed on plants or insects. They compete with, rather than eat, each other. But the gap in their ecology left by the absence of any large carnivore is one that surviving reptiles have not been slow to fill. Australasia abounds in snakes, crocodiles, and lizards, all eating marsupials as fast as they can—and none is more successful in this endeavor than a group known as the monitors.

These are opportunist lizards that can walk, run, swim, dig, or climb with equal facility and will eat absolutely anything. Their average length is less than a meter, but in the warmth of the equatorial Sundas, in the presence of large easy meals promised by flourishing populations of pigs and deer, one group of monitors has gone berserk.

On Flores, Padar, and Rintja islands, but especially on the remote volcanic outcrop of Komodo, the monitors have grown and grown until they are now the largest living lizards in the world. At over 3 meters in length, with a weight sometimes in excess of 130 kilograms, they are easily the biggest reptiles to have stalked dry land since the Age of Dinosaurs.

Dragons are still alive and well and living in Indonesia.

Word of their well-being, however, has been slow to spread.

The first account in the literature is that of a pioneer aviator who made a forced landing on Komodo in 1910. To add to his already considerable difficulties with primitive equipment, he found to his dismay that he had also to contend with enormous reptiles that prowled hungrily around the wreckage of his plane.

Nobody, of course, believed a word of his story until an officer of the Dutch Colonial Infantry on expedition to Komodo succeeded in shooting two of the animals. These specimens were taken to Java, where they were studied by a Dutch zoologist named Ouwens who, in 1912, finally reconciled folk legend, myth, and science by publishing the first complete description of a living dragon—which he correctly identified as an exaggerated monitor lizard and christened *Varanus komodoensis.*

At that time Java and all the islands east of it, including a large portion of New Guinea, were part of the Dutch East Indies. Colonial administration, however, was spread very thin and the running of the remote regions was left in the hands of traditional rulers or sultans.

Komodo fell under the jurisdiction of the sultan of Bima, who lived on the large island of Sumbawa to the west. For generations, this ruler and his predecessors solved all their penal problems in a very simple, almost classical, way. They deported offenders to Komodo. Which was a punishment that amounted, in effect, to feeding them to the dragons.

So when the reptiles moved out of the realms of myth into scientific print, and zoos and museums and private collectors began to take notice, the sultan of Bima took action in defense of tradition. In 1915, he introduced the area's first conservation legislation—a law putting the dragons, known locally as *ora,* under complete protection.

And it worked. In 1927, the Dutch Resident on Timor was persuaded to honor custom by passing an ordinance that imposed a mandatory fine of 250 florins on anyone found disturbing the dragons in any way. In 1928, Komodo was officially proclaimed a wilderness area accessible only by special permit, and when Indonesia finally became independent in 1945, steps were taken to establish the island as one of the new nation's first national parks.

I didn't get to Komodo until 1965, but in the years between 1910 and my arrival, very little had happened to disturb the island. A small number of official expeditions had come and gone, taking with them about eighty dragons for foreign collections. The biggest change was the growth of a substantial village of stilted huts with palm thatching sprawled along the shore of a southern bay near the only reliable source of fresh water.

The deportees and their descendents now numbered several hundred, far too many for the dragons to deal with, and they had established an uneasy working relationship with their erstwhile executioners. The area around the village had been cleared of vegetation, not so much for cultivation as to deprive the dragons of natural cover, and a series of staked barriers had been erected as additional defenses. No goat was ever left tethered and all graves had to be piled high with cairns of stone to prevent the dragons from exhuming recent burials.

The whole arrangement, and the ever-present air of faint menace, was strongly reminiscent of that other oceanic island where a similar community had lain behind its fortifications, hardly daring to sleep, waiting for the next awful visit of King Kong.

My own first sight of a dragon was almost disappointing. The usual way of showing them to visitors is to kill a goat and hang the carcass in a tree near a well-known lair. Komodo is hot and it is not long before the sickly sweet smell of putrefaction brings dragons out to investigate.

For three hours I lay on the edge of a dry ravine peering down at the bait, listening to the buzzing of the flies and the occasional raucous call of a white cockatoo. Nothing happened except that a tiny *ora,* less than 60 centimeters long, far too small to be dignified as a dragon, scrambled along the tree trunk in an attempt to get at the goat from above. He looked to me like any other lizard and I was half prepared to discount the stories I had heard of the predatory adults, when one of them finally arrived.

There was a rustle of dry leaves a little way up the shady ravine and then, without further warning, a prehistoric monster stepped out into the sun. No photograph or film can capture the threat of a dragon in the flesh, in search of food. He stood with his broad head held high, tasting the air with a forked yellow tongue, looking, it seemed, directly at me with a dark cold basilisk eye. His dusky scarred skin was heavily scaled, hung in folds about his powerful neck, and was draped loosely over a body like a great tree trunk propped up clear of the ground by four muscular legs with long clawed toes.

When he moved, it was with surprising grace, throwing his enormous weight from side to side with a fluid reptilian movement of the hips. Reaching the suspended bait, he lunged up and, taking hold of the goat's haunch in his teeth, ripped it clear of the branch with one sweep of his head.

He paused for a moment as a cloud of flies and dust swirled about him before disemboweling his prey by biting open the distended belly, slinging the viscera violently from side to side to scatter their contents as he gulped the intestines down. Then, with a loop of mesentery still hanging from the corner of his jaws, he turned his attention to the rest.

One bite with the serrated, sharklike teeth and a few raking movements that threw the carcass back over his shoulder were sufficient to saw through the spine and separate the hindquarters from the head. Then both pieces were swallowed whole. The complete disposal of the 20-kilogram goat took little more than ten minutes for an animal used to consuming up to 80 percent of its own body weight at a single sitting. All that was left in the end was trampled sand and a few stray wisps of hair.

I was impressed.

Dragons hear only moderately well and their sight (at least as far as movement is concerned) is fair, but their sense of smell is acute. I once saw a dragon follow day-old deer tracks along a sandy beach, testing each one with its tongue, moving always in the right direction. It is possible that they locate and capture injured, ill, or unwary animals in this way, but by far the largest part of their prey is taken by surprise from ambush.

Wild pigs are grabbed by the legs and eviscerated as they fall. Deer are often taken by their rumps, away from sharp hooves and antlers, and borne down to the ground by the sheer weight of the attacker. And even water buffalo and feral horses, which have been introduced to the islands in the last century, are not immune. I know of at least one buffalo that cannot have weighed less than 600 kilograms being killed by a 2-meter dragon.

Despite these abilities, I discovered after some time on Komodo that dragons I met on the trails would usually give way to me and that I could approach to within a few meters of one that was already feeding. I was always careful, but must admit that I became a little blasé and began to think of the dragons more as scavengers than active predators. At least until I met 34-W.

The only intensive study yet made on Komodo was one undertaken by Walter Auffenberg of the Florida State Museum. He spent over a year on the island and numbered all the adult dragons that he saw, identifying individuals by differences in the markings on their snouts. He had to resort initially to a complex catalogue of distinguishing marks, but there was one animal he never had trouble recognizing.

Of this one, he says, "When I met him on a game trail, this individual would approach with mouth open and tail bowed in threat. He once stalked my children on the beach and often came directly into our camp. On one occasion he drove us out of our tent, stuck his head into a knapsack, removed a shirt and tore it to shreds. There was no question in our minds that if given the opportunity this animal would attack, whether provoked or not."

I first came across Auffenberg's 34-W on a forest path with thick bush on either side. I waited for him to give way, but his only response was to lower his head, puff out his throat, and hiss. I backed off, but he kept on coming, walking stiff-legged, body held high off the ground. So I turned and ran.

Not long afterward, Sugito Tirtowono—an Indonesian wildlife officer from the Directorate of Nature Conservation in Bogor—was attacked by 34-W and succeeded in driving him off with stones, but others have not been so lucky.

A man from the Komodo village was working in a forest clearing with his two sons, trimming sticks of wood into equal lengths, when 34-W appeared. They jumped to their feet and fled. But the youngest boy, just fourteen years old, ran into a low hanging vine and was hit from behind by the dragon before he could disentangle himself. The other two returned and succeeded in driving the animal off with sticks, but it had already torn huge chunks of flesh from the boy's back and he died within half an hour.

On another recent occasion, four poachers from Sumbawa came to Komodo to hunt for buffalo. While camping in the hills, one of them became ill and the others went to the village for aid. By the time they returned, there was nothing left of the individual but a leg.

These are terrible tales, but there is something strangely reassuring about them. They give me the same feeling of guilty pleasure I get from hearing that a Christian missionary has been eaten by cannibals. It is nice to know that such things can still happen.

Komodo is one of the most beautiful places on earth. Its high sculpted peaks, crowned with silhouettes of *lontar* palm, sweep down to grassy plains covered in acacia parkland that is always so superbly lit that it gives one the impression of walking through one of those elegant formal dioramas in a natural history museum. Orioles, drongos, and jungle fowl in the coastal bush compete for attention with whales, dolphins, and manta rays that glide through the deep still water near the shore.

It is a Garden of Eden, complete with resident reptile, and I feared greatly for the future of both when cruise ships began to call there in 1973.

The visits were sporadic and well organized, with guides and villagers doing everything possible to limit the impact of tourists on the island. But tourism is an insidious thing, far more tempting and potentially more destructive than any forbidden fruit. I didn't think that the ecology could stand it. But I had reckoned without the dragons.

On July 18, 1974, a German group came to Komodo. They set out on foot to visit a site where goat baits had been placed to lure some of the larger dragons within range of the usual battery of lenses. It was a hot dry day and one of the party—a seventy-eight-year-old Swiss baron—found the going too hard for him. He stopped to rest in the shade of a tree and when the group returned two hours later, there was nothing left of him but a Hasselblad camera with a broken strap.

Search parties combed the island for two days without finding any further trace of the old man. The Indonesian authorities, understandably anxious not to rock the cruise boat business, did everything possible to play down the dragons' role in his probable demise, and Baron Rudolf von Reding Biberegg was officially described simply as having "disappeared."

Access to and movement on Komodo was stopped for a while, but by 1976 there were once again parties padding about the island almost every week.

Last year I returned to Komodo myself, fearing the worst, expecting to find Coca-Cola stands on the beach and Kodachrome wrappers littered all along the trails. But I was wrong. The island is as lovely as ever, and sitting round a fire in the village talking over the sights and sounds of the day with the local people, I discovered why.

Just as the tourist boom was showing signs of recovery, the dragons had struck again, this time at a young and athletic Frenchman—leaving nothing but a blood-stained Adidas running shoe.

So it is still going on. Dragons are still eating and frightening people. At least often enough to keep the old myths alive.

I think it is important, but I also believe that our relationship with dragons goes back so far that we might not even need periodic reminders of this kind. I suggest that it all really began 65 million years ago when our distant ancestors, the first small mammals, were being hunted by the last of the great dinosaurs—and that there is a part of our brains in which the memory of those antique chases lingers still.

Paul Maclean at the University of Toronto has pointed out that we have not one brain but three. And that these differ from each other in function, anatomy, and biochemistry. He identifies an advanced brain or neocortex, which is present in any quantity only in higher animals such as carnivores, whales, primates, and people. Beneath this "thinking cap" lies an older brain, the limbic system, which governs some of the stronger emotions that we feel in common with many birds and lower mammals. And underneath both these brains, like a swelling on the end of the spinal cord, is the reptilian complex—something that has survived, almost unchanged, for several hundred million years.

We think with the top brain and feel with the middle one, but still seem to be subject—particularly when asleep or when consciousness is being held at bay—to rumblings and stirrings from the lower levels. The dragons live on, not only on Komodo, but perhaps also in the darkness that lurks deep inside each of our heads.

Jung was impressed with the number of folktales, mainly from the West, in which archetypal heroes take on dragons on behalf of their beleaguered people. He saw the stories as evidence of the danger we all feel of our newfound consciousness being swallowed up again by the older, more instinctive, unconscious. The hero's task is to overcome the monster of darkness and let in the light. The killing of the dragon is symbolic of the hero, the self, coming to full awareness.

There is some merit in this largely Western interpretation, but there are dangers too. By identifying the dragon solely with evil and the forces of darkness, we run the risk of missing the point. The Oriental view of dragons as symbols of power, as creatures of the Storm that can be forces for both good or evil, is perhaps closer to the truth. We need to subdue the dragons, but it would be suicidal to banish them altogether. To do so would be tantamount to destroying part of ourselves—a part that still has a useful purpose.

The best dragon tales are those in which the hero succeeds not only in defeating the dragon, but in freeing a helpless maiden that the dragon holds captive and is about to devour. In modern versions it is usually the handsome cowboy who rescues a girl tied to the railway tracks in the path of a fire-breathing iron engine.

These myths show our new conscious selves coming to terms with the old unconscious areas and in the process releasing our feminine counterparts—those areas of personality that are gentle and creative, fertilizing the earth.

We need our dragons. They remind us of our origins and our incompleteness. Long may they live in dreams and on desert islands. Even if it costs us an occasional Swiss baron.

7

The Arteries of Earth

In 1964, the Swaziland Iron Ore Development Company completed negotiations with the government of Japan. These concentrated on a mountain known locally as Emabomvini—"The Place of the Red"—a peak of almost solid hematite long sacred to the Swazi people.

Ten years later, nothing was left but a massive hole in the ground. A ridge over a kilometer high had been bodily removed and 50 million tons of rich iron ore was ferried over the ocean. Crews on the bulk carriers amused themselves on the long voyage to the East by sorting through the soft red rock and picking out pieces of darker dolerite. There were plenty of these and every single one proved to be an artifact—a prehistoric stone tool shaped for a purpose by early human hands.

The purpose at first was obscure. Stone Age implements abound in Africa. There are places where the whole surface of the soil is carpeted with flakes and cores and beautifully shaped hand-axes. Archaeologists in Swaziland had been aware of the tools on Emabomvini. Thousands of them lay scattered all over the sacred mountain and on other peaks nearby. But it was not until the bulldozers moved in that anyone realized the extent of the ancient industry. Doleritic fragments were found in patches of disturbed earth as much as 30 meters underground, and a last-minute expedition was organized to rescue evidence from the heart of the mountain before it disappeared altogether. This produced some astonishing results.

Most of the chunks of dolerite, which is foreign to the area, had been chipped and flaked into shapes reminiscent of picks, hoes, and chisels. And many of them were buried beneath thousands of tons of red iron oxide, as though they had been used and then abandoned at the ends of deep tunnels. It looked very much as though people who had no iron themselves were using stone tools to mine iron ore long before the beginning of the Iron Age. This, in itself, was surprising enough. But what really shook the archaeologists was the discovery in one of the tunnels of charcoal nodules that could be accurately dated by radiocarbon techniques. These proved to be forty-five thousand years old.

Until very recently, mining was supposed to have begun only seven thousand years ago. At that time, Neolithic communities in Europe were just beginning to clear the forests to make way for agriculture. They needed tools tough enough to fell the hard woods of oak and elm, and dug deep holes to get to the necessary flints. Africa was regarded in archaeological circles as a cultural cul-de-sac, to which such modern practices had only been introduced by Iron Age immigrants during the last few thousand years. But suddenly, way down south in Swaziland, here was evidence of mining—which is a highly sophisticated activity—at least six times older than anything known in Europe.

And there was worse to come. Later discoveries in another part of Swaziland produced proof of underground activity taking place over a hundred thousand years ago. And in every case, the ancient mines were not only excavated by hand with considerable skill, but had been laboriously filled in again afterward with thousands of tons of stone. Our whole Stone Age chronology has now had to be radically revised. Southern Africa, as a result, begins to look less like a backwater and more like the center of technological invention. A place in which innovation thrived while Europe and Asia were virtually stagnant, waiting for the agricultural revolution to begin.

Africa's Stone Age miners were, it seems, after precisely the same stuff as those who, in the last decade, knocked the mountain down. But they needed it for a different reason and, unlike us, they treated it and its source with elaborate respect.

The ancient miners scooped out the bright red hematite for use as a spiritual cosmetic. It was, and in many areas still is, a symbolic surrogate for blood. Powdered red ocher is bartered and traded, used to decorate the living and to anoint the dead. It is a powerful cultural catalyst, painted onto ritual objects and fired into pots for ceremonies that cannot take place without it. Of all minerals, it is perhaps the most sacred, playing a vital role in the myths and mysteries of ancient metallurgy—and in the performance of more modern rites. It has always been seen and known as the Blood of the Earth. And the task of tapping it was never taken lightly.

Mining is, in essence, a violation of Mother Earth. It is a very bold undertaking and many people still refuse to do it. "You cannot ask me," said a Sioux medicine man, "to dig in the earth. Am I to take a knife and plunge it into the breast of my mother? Must I mutilate her flesh to get at her bones?" Wherever need overcame such natural reluctance and mining did take place, it was always surrounded by elaborate rites and superstitions. Digging never began without appropriate sacrifices and was accompanied by the appeasement of daily offerings. And, most important of all, when the work was done, when the miners had what they needed, the damage was repaired by filling in all shafts with the rubble taken from them. The skirts of the violated earth had to be decently rearranged.

All the old Swaziland diggings, the oldest known mines in the world, were meticulously repaired. The holes were filled and the wounds allowed to heal. Only in this way, it seems, could the diggers live with their sacrilege and keep faith with the earth itself.

It is hard for most of us, who have lost direct touch with earth, to understand things like this. We may know the myth of Antaeus—the son of Poseidon who defeated all comers until Hercules killed the giant simply by holding him aloft, out of contact with the earth, so that his strength just ebbed away. But we fail to see how it applies to us. It does. We spend our city lives in suspense, perched on plastic, propped up by scaffolds of glass and steel, slowly being drained of natural energy.

It need not be this way. We seem to be stuck with many of the new materials, but there are ways of arranging them that are very much more sympathetic and could put us back in tune. All it takes is a little sensitivity.

All over the world there are special places that, by general agreement, are more inspiring than others. The sanctuary at Delphi in Greece, the base of Ayre's Rock in Australia, the top of Glastonbury Tor in England, the shrine at Ise in Japan are some of these. They have little or nothing in common, except for their qualities of grace and mystery. All have been holy places—recognized by shamans and wizards, visited by bards and witches, guarded by priests and hermits. They have been decorated with charms and embellished with temples, and go on being recognized, doing something basic and biological that makes them valuable to people of all kinds in every age.

Some of these sacred centers stand alone, but often they are linked with others, setting up patterns in the landscape.

The deserts of Australia are laced together by traditional tracks connecting centers at which superbeings are said to have performed creative acts during a mythical past known as the dreamtime. Aborigines believe that every plant and animal possesses *kurunba,* a life essence that has to be renewed. And that it is man's responsibility to suck spirit energy out of the rocks and keep this essence alive. So, every season a group of worshipers travel along the old paths, stopping at the sacred centers to perform the appropriate rituals and help stimulate the flow. Many of the landmarks on these routes are well known and easily identifiable from old descriptions in song and dance. Young initiates can go straight out into the desert for the first time and find the water holes, trees, and rocks involved. For them, each special natural object, each sacred place, is part of the body of an ancient being, and the sight and recognition of it brings deep satisfaction. It plugs them directly back into the system, no matter how long they have been away.

A similar relationship exists among megalithic sites in England at Stonehenge, Walker's Hill, Silbury Hill, and Avebury—which were once connected by serpentine avenues and tracks marked by the erection of "motherstones." The summit of Wilcrick Hill in Wales forms the center of a magnificent geometric spiral of earthworks that covers huge areas of the old county of Monmouthshire. Christian shrines in tenth-century Bohemia were constructed on a series of holy hills along recognized routes of pilgrimage. The Andean cities of Peru, including the spectacular Machu Picchu, lie at the sites of special shrines along the line of the old Inca High Way.

None of these routes or circuits are arbitrary. All act like circulatory systems, tying the body of a landscape together in ways that continue to be meaningful. The aborigines are by no means alone in their ritual walkabouts. They are joined every year by thousands of British parishioners who take part in a ceremony known as beating the bounds. On the appointed day, the people of a parish gather together to comply with an injunction made four hundred years ago by Queen Elizabeth I, who said, "The people shall once a year with their curate walk about the parish as they were accustomed. . . . The curate in certain convenient places shall admonish the people to give God thanks."

These "convenient places" are all on the sites of ancient markers—trees, standing stones, shrines, or springs. And at each one the parishioners pause, not only to be admonished, but also to beat the site with willow wands in order to reenergize it and stimulate its vital flow. These ceremonies in Britain can be traced through Elizabethan times back at least to the Roman occupation and the rites of Ambarvalia—in which priests led the people in procession to visit holy groves, stones, and altars, at each of which a sacrifice was made. The Romans used sheep or pigs, but they may in their turn simply have been perpetuating even more ancient Celtic customs that required something stronger, usually the sacrifice of a child. When "beating the bounds" today, the parishioners still make symbolic sacrifices, laying a child in a cross-shaped indentation in the soil or holding a young boy, a different one at each site, upside down and bumping his head on the holy ground.

Tacitus, the Roman historian, tells of sacrifices being made each year in sacred groves identified by priests who "can feel the presence of the goddess in these Holy of Holies and attend her there." Every society has such sensitive people who not only keep the memory of the old sacred sites alive, but are adept at finding new ones. Wherever holy places exist, they have a history. They are accompanied by legends that tell of their discovery. They are found because of a dream or a vision; they come to light as a result of some divine omen; or they are simply pinpointed by someone who felt strongly about them—someone who divined (and this particular verb is very appropriate) the exact spot.

A diviner is someone who discovers the unknown by intuition. He is usually a priest or holy man—someone who is familiar with the necessary rituals, but has also an added ability to discover those places that are most auspicious for religious purposes. All over the world, the methods are much the same.

Some diviners study the heavens and relate these to features in the local landscape, interpreting natural highlights, trees and rocks, hills and valleys, in terms of constellations and planetary attitudes. They work as astrologers still do, calculating first on the basis of fixed relationships and then applying the results intuitively to the particular situation.

Others use simple mechanical aids, such as compasses and sticks, feeling their way to the right answers and the proper places, walking through the fields in search of appropriate inspiration. The dowsers who find water or minerals, lost rings or missing people, are part of the same tradition, responding perhaps to similar signals. I know from my own experience with such people in the field that this talent is a real one, capable of critical test, but I am a little bewildered by the hardware. Willow wands, whalebones, coat hangers, amber beads, brass rods, and surgical scissors have little or nothing in common. I suspect that all are in the end unnecessary. They are aids that we can do without. Some of the best dowsers work with their bare hands, picking up the messages, listening to fluids flowing in the arteries of earth.

However they work, there is little doubt about the results. There is something in the chosen holy places that makes them very different, obviously different, from other places simply picked at random. The old sacred spots have a mood and a balance that are unmistakable. They tend to occupy positions of advantage. These need not be prominent in the sense that they straddle the highest hills, but they always lie at the focal point of an area. At a point where physical and spiritual forms seem to combine to produce a kind of emotional equilibrium. By comparison, new sites of community attention that are allocated by town planners or municipal architects on the basis only of space available are awkward and barren locations, many of them already empty and forlorn.

I am convinced that the special feel of holy places is real and that in responding to them we are not just being carried along by their reputation. Other species seem to share our feeling for them. It is true that swarms of insects often congregate over obvious landmarks—many trees in summer carry an umbrella of flies that can be seen even from a distance. But church steeples and temple spires seem to exert a particularly magnetic attraction, bringing in so many bees, wasps, and midges that they appear to be generating plumes of smoke. Cattle too tend to congregate in similar ways about standing stones or old earth mounds and barrows—almost as though they felt the same attraction.

Perhaps they do. Wild antelope and deer select resting places with great care and attention, but the spots they settle on seem very often to be less secure than others only a short distance away. Cattle in a field choose to lie down in one particular area despite the fact that it offers in many cases no apparent advantage in the form of food, shade, or shelter. These tendencies were not lost on those intent on finding holy places—and were often put to good use.

The Celtic monks of Lindisfarne, wandering about the north of England with the body of their founder, Saint Cuthbert, were led to his eventual burial site by a stray cow. The spot is now covered by Durham Cathedral. Waltham Abbey, another of England's major shrines, was located by allowing a team of eight oxen to roam cross-country until they came to a stop of their own accord. In parts of Africa and North America, places of similar reverence were sometimes established by sewing together the eyelids of an animal and letting it walk blind until it chose, with its other senses, a place secure enough to lie down.

It is hard to say what it may be that influences such decisions. We know from recent work on homing and migration that animals are capable of responding to magnetic fields, ultrasound, infrared, and an extraordinary variety of other subtle sensory stimuli. It is likely that we, if only unconsciously, still respond to many of the same signals. As nomadic hunter-gatherers, we may once have found them vital for our survival.

To begin with, we simply used the earth, fitting our lives to its easy rhythms. Any alteration to the natural patterns was unthinkable and unnecessary. We were simply carried to and fro by the ebb and flow of the seasons, finding ourselves—as migrant animals still do—returning to favored places at regular intervals. The choice of such places had a great deal to do with the availability of food and shelter, but right from the start there were other considerations. There is something else that we, and many other living things, appear to need almost as much. We seem to have an awareness of, and a hunger for, essential harmony.

It has always been so. We all have earth's measure and, given the space to express this heritage, we easily and inevitably drift toward those places on the surface of the planet where things are in kilter. It is here that we can sit in comfort and sleep in peace. Here that we find the freedom to dream and enjoy the sense of belonging to something bigger. It is in these places that we made our first worshipful marks, decorating rocks or trees in ways that turned them into shrines. And it was around these early altars that there grew the first simple temples. And over the ruins of these that we have now built the monuments to our current gods.

The succession is clear. There is a continuity and flow in these things that creates the best traditional landscapes. Such things look right and feel good because they fit. The best buildings grow where historical consensus agrees that they should. These things have to evolve naturally. They come into being as a result of a sort of spiritual engineering that takes place between the earth and those who live close enough to it to feel the rhythm of its breath. Under such conditions, a new structure achieves a form whose lines enhance and enrich, rather than violate, the character of a landscape. And the roads that lead to such places of importance are paths with heart—smooth and vital arteries instead of ugly varicose veins.

It is hard for even the most sensitive architect to contrive anything of the sort. But there is at least one book of rules, one set of ancient instructions, to help him do so.

The Chinese have always recognized a magical link between man and the landscape. They see the world and themselves as part of a sacred metabolic system. Everything in it pulses with life and each thing depends on everything else. Nature reacts to change and that reaction resounds in man. When the earth is healthy and prospers, we thrive. When the balance is destroyed, we suffer. Our fate is inextricably bound up with the cycles of heaven and earth and with the flow of the weather. And at the root of all things lie the twin forces of wind and water—*feng* and *shui.* So all health and prosperity depend on the maintenance of harmony between these forces. On the science and art of *feng-shui.*

At the heart of feng-shui lies the concept of *ch'i,* the motivating principle that links spirit and substance. Ch'i inflates the earth, moves wind and water, and breathes life into plants and animals. It pulses through the planet, motivating each thing that moves and irrigating every landscape. Where it flows smoothly, there is a touch of magic light, something that can be understood intuitively, but never fully conveyed in words. At its best it brings the sort of harmony that, in the words of a seventeenth-century devotee, makes a place in which "the hills are fair, the waters fine, the sun handsome, the breeze mild; and the sky has a new light: another world. Amid confusion, peace; amid peace, a festive air. Upon coming into its presence, one's eyes are opened; if one sits or lies, one's heart is joyful. Here ch'i gathers, and the essence collects. Light shines in the middle, and magic goes out on all sides."

The laws of feng-shui are complex, but in principle assume the existence in earth of two complementary currents—one male, light, and active; the other female, dark, and passive. The first is identified with the blue dragon, and it and its attributes in a perfect landscape ought to be found on the left-hand side. The second is the white tiger and should ideally be found on the right. Hills and pillars of rock are naturally male areas, and softly undulating ground, female. The ideal proportions of each are said to be 3:2 and the most propitious places in such mixed landscapes are thought to lie in the area of transition, where the two currents meet, preferably at a steep angle.

Given free access to a wide area, such places can be found. The site of the city of Guangzhou (Canton) at the head of its horseshoe bay is one, but even without such freedom of choice, all is not lost. Even a tiny plot, hedged in among others of equally monstrous geometry, can be enhanced by the careful construction and adjustment of appropriate furrows and stones. The task of the professional is to define the difficulties and to suggest, in accordance with the principles of feng-shui, an appropriate remedy. This might mean something as drastic as moving the entire house to face in a different direction—or something as trivial as planting a tree to one side of the door or hanging a mirror in an unbalanced hallway.

The best feng-shui practitioners are part ecologist, part musician, and part priest. They decode an environment, lend an ear to its discords, and put man and his works back into tune. Postrevolutionary China is prey to all the disharmony that seems inevitable when cities and industries arise and are allowed to grow out of control. But houses, farms, and villages throughout much of rural China still bear powerful testimony to the effectiveness of feng-shui. There is a beauty in siting and setting, an effortless ease in landscaping, that wonderfully reconciles the needs of people with the subtle energies of their surroundings. It really works.

How it does is another matter.

In 1983, a team of physicists at the University of Rome set up a complex piece of apparatus designed to detect gravity. This consists of huge blocks of solid aluminum so delicately balanced that sets of instruments are able to monitor any unusual vibration in their component electrons. One block is in Italy; the other, to guard against the possibility of local error, stands 600 kilometers away in Switzerland—the theory being that anything recorded simultaneously in both blocks is likely to be due to more distant and more basic causes. The experiment has failed so far to produce any evidence of gravity waves, which remain purely hypothetical, but the vast two-pronged aluminum antenna has picked up something quite unexpected.

Precisely every 12 sidereal hours—that is, hours set relative to the earth's movement through the galaxy, and not about its own axis—the earth throbs. The cause remains mysterious—it may be something to do with lunar and solar tides—but the effect is clear. The pulse of our planet beats regularly, once every 11 hours, 57 minutes, and 57.3 seconds, earth time. And its effect can be felt more strongly in Rome than in Geneva.

Is this ch'i, bubbling up through cracks in the planet? Probably not. But it is evidence at least of the existence of a global rhythm that has local peculiarities.

Among the great triumphs of twentieth-century science are the satellite surveys of earth's surface. And of all these, none has been more revealing than the Seasat mission flown by NASA in the summer of 1978. Circling the globe at a height of 800 kilometers, this satellite swept the surface of the sea, measuring its level to the nearest 10 centimeters. The result is a marvelously detailed map of the world ocean, showing that it is not level at all. Gravitational anomalies produced by the variable thickness of earth's crust mean that there are differences of as much as 100 meters in the mean level at various places. There is a trough, almost a hole, south of India and another great dent in the area that has come to be known as the Bermuda Triangle. And there are equally permanent hills around Iceland and Hawaii, and in the Sea of Japan.

When the data are expanded to include the land areas, it is clear that our globe is a confusion of gravitational peaks and troughs that are relatively independent of general geography. Positive anomalies, places of unusually high gravity, bear some relation to the great mountain ranges. There are gravity peaks that roughly coincide with the Rockies, Andes, Himalayas, and Alps—all areas of thrusting geology and active, masculine mountain-building. The domains of the blue dragon. And there are negative anomalies, places where gravity is unusually low, along Africa's Great Rift Valley, in the Amazon Basin, and around Hudson Bay. All places of subsidence, the sort of cleft, yielding, feminine landscapes in which one would expect to find white tigers.

It would be wrong to stretch the analogy too far. The forces of ch'i are probably at least partly psychic and aesthetic. But it is interesting that we have begun to be able to measure and identify characteristics of earth that are rooted in areas far below the surface. And that the little hills and valleys that concern feng-shui are valid superficial symptoms of the great tectonic tensions that move continents around like curds in a bowl of whey. It would be equally wrong, and patently unscientific, to ignore the possibility that some of us, perhaps even all of us, are as sensitive to these tensions as any passing satellite.

An awareness of natural rhythm is not confined to African miners or Chinese geomancers. It is part of our birthright. A gift from Mother. With one ear to the ground, you can almost hear her great heart beating. It is a very comforting sound.

8

The Currents of Life

In fable, the salamander is said to be a sort of lizard that lives in fire, whose heat it overcomes by the chill of its own body. In fact, it is a small amphibian, a little like a lizard, which is found in cool mountain streams. The Roman scholar Pliny, ever curious, tested the myth by putting one to the flame and reported, somewhat sadly, that it was burned to a cinder. Nineteen hundred years later, a professor of anatomy at Yale set up an experiment that was not only more humane, but also so successful that the salamander deserves to be remembered, along with Newton's apple, as one of the great sources of scientific inspiration.

In 1935, Harold Saxton Burr was beginning to be intrigued by the electrical characteristics of living systems. He was aware that some animals, such as electric eels, could produce currents powerful enough to stun their prey. But he suspected that all creatures, even the simplest ones, were probably surrounded by some sort of electrical field—and he proved it with one of the most simple and elegant biological experiments ever made.

Burr took a salamander and turned it into a dynamo. In its most rudimentary form, a dynamo consists of an armature (usually a loop of copper wire) that is rotated (by falling water or passing wind) inside a magnetic field so that it makes and breaks that field in rapid alternation, producing an electric current. Burr assumed that a living salamander was producing an appropriate field. So he floated one in a dish of salt water, which conducts electricity almost as well as copper wire. To complete the circuit and obtain an armature, all he had to do was immerse two electrodes in the water and attach them to a galvanometer sensitive enough to measure a small electric charge. And then he spun the salamander.

The result was dramatic and convincing. As the salamander turned, or as the dish moved round the little amphibian, the needle on the galvanometer was deflected first to the left and then to the right in the regular negative-and-positive pattern of a perfect alternating current. And all he had to do to stop the current from flowing was to lift the salamander out of the water. Without the animal in the circuit, no current at all was produced.

Burr was not, of course, the first to suspect that life was electric. The instrument he used to measure salamander-power was named after the eighteenth-century Italian anatomist Luigi Galvani. And it was another amphibian, the common frog used as a model in his dissection classes, that started Galvani thinking along similar lines. He noticed that the muscles of even an amputated frog's leg twitched when brought into contact with certain metals, and suggested that electricity must be involved. But where did it come from? Being an anatomist, with a natural predilection for living things, Galvani decided on the muscle rather than the metal and proposed the existence of "animal electricity." He and a brilliant young German naturalist, the baron von Humboldt, began to experiment with the electrical nature of muscles and nerves. Together, they produced some superb analytic work, but in 1794 it was all called into question by a discovery made by another Italian.

The physicist Alessandro Volta was equally intrigued by the high kicks of Galvani's dead frogs. He was not very keen on dissection and tried instead putting an assortment of different metals together, until he discovered that the contact between copper and zinc was sufficient on its own to produce an electric current without any involvement of living tissue. A furious debate began between Galvani and Humboldt on one side, and Volta and the French physicist Charles Coulomb on the other. The argument raged for years until Volta clinched the victory for his side by the invention of an independent inorganic electric battery that was capable of storing electricity. This was widely seen as a triumph and Galvani was considered to have been completely discredited. He was not completely forgotten—the steady flow of current set up by two metals in contact came eventually to be known as galvanic electricity, and we still refer to someone who is provoked into sudden action as being "galvanized"—but Galvani himself died disillusioned and embittered. And "animal electricity" was all but ignored.

It took a salamander to get it going again in the twentieth century. Having proved that even such a small, slow-moving animal had its own electric field, Burr at Yale went on to work on trees and human beings. He tied a voltmeter to a magnificent old maple tree in Connecticut and kept it wired up for 20 years of continuous recording. The records show that the tree's field fluctuated in response to a 24-hour diurnal rhythm, to a 25-hour lunar rhythm, to a longer lunar cycle that reached its peak as the full moon passed directly overhead every 29.5 days, and to a pattern of sunspot activity that rose to a maximum roughly once every 11 years.

This variation in electric potential in accordance with external factors occurs also in humans. We show the same circadian fluctuation as a tree, but also demonstrate even more marked changes as a result of internal forces. There is a substantial voltage rise, lasting about twenty-four hours once every month, in the electrical field strength of most women—taking place precisely as they ovulate in the middle of their menstrual cycle. It has long been known that basal body temperature is about 0.3° C higher just after ovulation—a fact that helps those who keep daily records to achieve or avoid conception. Now research at King's College medical school in London reveals that this general temperature rise is linked with the ability of peripheral blood vessels to respond to local changes in temperature, making extremities warmer at this time than any other. And making physiological sense of an ancient male suspicion that women's hands and feet tend to be uncomfortably cold at most other times, when blood volume and electric field strength are relatively low.

Burr was never discredited in the way that Galvani was. Following the publication of a paper entitled "The Electrodynamic Theory of Life" in 1935, he continued with his research—showing that there were bioelectric correlates of wound healing and certain kinds of disease. He measured and identified the field properties of developing eggs and embryos, suggesting that it might be these that governed differentiation and growth. And he discovered variations in the electrical properties of seeds of hybrid varieties of corn and cotton. He published another sixty papers on bioelectricity before retiring in 1956, but despite twenty years of careful and methodical work, he seems now to have been almost completely forgotten.

There is something about the whole question of bioelectricity that continues to frighten science. The haste with which Galvani's early work was set aside in favor of Volta's gadgetry was almost unseemly—and seems to persist. Volta was right about direct electric currents being created by the reaction between metals, but Galvani was justified too in assuming that his frogs had potential of their own. Burr has proved the point and yet suspicion lingers on. Animal electricity continues to be set aside (along with "animal magnetism") as an antique conceit, in favor of the distinctly Voltaic mechanics of molecular biology. We are learning extraordinary things about biochemistry at a cellular level and remain woefully ignorant of the true nature of the whole plant or animal.

Advances in electrophysiology make it abundantly clear that, in all living systems so far examined, the membranes that surround cells maintain an unbalanced distribution of ions. There is a difference between the charge inside and outside a cell and an electric gradient, usually a drop in voltage, across the membrane. Nerve cells make good use of this potential difference to transmit information over long distances. But they are not the only ones to do so.

It is very difficult to measure the voltage in a single small cell, but Lionel Jaffe at Purdue University in Indiana solved the problem years ago by setting up a beautiful little experiment with a number of cells of *Focus*—a common brown seaweed usually known as wrack. He took egg cells from the plant and lined them up in a narrow tube with electrodes at either end. Then he shone a light down one end of the tube to persuade the embryos within to grow. Long before they could, each of the cells polarized, becoming positively charged at one end and negative at the other—and all did so in the same way, with their positive poles nearest the light, so that tiny voltages in each added up along the length of the tube, arranging the cells like a row of batteries in series, producing a current big enough to be measured without difficulty. It looks as though the separate cell charges in most multicellular organisms add up in just this sort of way to produce an appreciable field.

The evidence also suggests that all embryos build up complex and distinctive fields in the same way, and that Burr was right. It could be these fields that guide their growth, giving every cell an address. Telling the cell, in effect, where and who it is and which of the genes in its enormous bank of information it would be appropriate to use. This, it seems, is how one part of an embryo "knows," for instance, that its task is to grow into a leg rather than an arm—although it contains all the instructions necessary to do both things equally well. If the future tail of a salamander embryo is cut off and grafted onto a position where the leg should be, it grows not into a tail, but into a leg. If you cut away half of a frog embryo at an early stage, it doesn't grow into half a frog, but a whole one.

In truth, we still haven't the faintest idea how this happens or what precisely it is that turns a fertilized egg into a frog or a flower. This remains the biggest mystery in biology, a subject in which Nobel Prizes are just waiting to be won. There are patterns of chemical information, of hormones and even vitamins, that seem to play their part in organization, but suspicion grows that the "master plan" that turns a single cell into the billion or more that go to make up a functional human being may be electrical. And, once again, it is the salamander that is providing the most useful clues.

Unlike us, a salamander is capable of growing its own spare parts. It can regenerate and replace lost legs, eyes, heart, spinal cord, and even parts of its brain. And at the State University of New York in Syracuse, the orthopedic surgeon Robert Becker has been trying to find out how this happens, in the hope that we too may be able to benefit.

Becker knew from Galvani's work two centuries earlier that electricity—a "current of injury"—was most strongly produced in any animal when the skin was broken. He suspected, as Burr had, that this flow of energy had something to do with healing and decided to compare the current in a salamander with that measured in a relative like the frog that is similar in size and complexity, but lacks the ability to regenerate its limbs. He amputated one foreleg from each animal and waited. For three days there was no difference. Both amphibians produced strong electrical potentials in their leg stumps and the skin on both began to heal, but on the fourth day, the salamander's current suddenly reversed its potential, turning abruptly from positive to negative—and radical changes took place beneath the skin.

The severed nerves in the salamander stump began to grow until they formed a new network in contact with the epithelial cells near the surface. This group of cells then changed character altogether, becoming once again unspecialized primitive growth cells of the kind that appear in embryos and are capable of going off in any direction, doing whatever is required of them. Within days, a new limb bud had appeared with rudimentary muscles fed by the appropriate nerves and blood vessels. And within three weeks, the missing limb had totally re-formed with every digit back in its proper place.

Becker was convinced that it was the change in charge that triggered this series of events in the salamander and he tried artificially altering the electrical potential in the stump of an amputated frog. It worked. The frog leg began to regenerate. He then turned to work with mammals and succeeded, by putting a small battery producing a constant negative charge in contact with the site of the injury, in getting the severed limb of a rat to start to grow again.

Bones in mammals repair themselves, knitting together after a fracture, but the regrowth of missing bone was unheard of until this experiment in 1972. Orthopedic surgeons all over the world began to sit up and take notice. Before long, patients in Japan, Europe, and the United States were having tiny hearing-aid batteries implanted close to severe fractures or alongside those that, in older people, showed a marked reluctance to heal. The results were dramatic. Electrical or electromagnetic osteogenesis rapidly became an accepted and successful practice for the treatment of nonunion in bones, but excitement seemed to be confined entirely to Becker and a small group of clinical orthopedists. The scientific establishment continued with its biochemical obsession and remained skeptical about the connection between life and any form of electricity.

It still is skeptical, but it is becoming increasingly difficult to ignore the growing flood of information on the effects of electric fields or currents on living things. In 1960, there were just three papers on the subject in the scientific literature. By 1984, there were over ten thousand—and many of these suggest that not all the effects are therapeutic.

At the University of Rochester in New York State, rats have been trained with food rewards to press a lever when they can detect an electrical field. They do this with considerable success, even when the field is very weak, but when given the choice, show that they prefer to avoid high-strength alternating currents operating at 50 to 60 cycles per second—which is the usual frequency for powerlines. Pigs seem to be even more sensitive, particularly when they are pregnant. And in one experiment with golden hamsters exposed to such a pattern, all the animals dismantled and moved their nests beyond the limits of the field within seventy-two hours. Those hamsters that had litters of young in their nests moved even more quickly, carrying their babies and the nest material to a less electric environment within twenty-four hours.

There is good evidence to show that humans have comparable sensitivity, complaining of a tingling or crawling sensation of the skin, and a stirring or prickling of fine hairs on the arms or in the nape of the neck. At low field strengths, this may be no more than inconvenient or unpleasant—workers in factories that make permanent magnets often complain of irritability, fatigue, and dizziness.

But prolonged exposure to powerful sources of electric energy can be lethal.

A study made in the British Midlands in 1981 produced some ominous results. Measurements were made there of the strength of the field produced by powerlines at a total of 590 houses in which suicides had taken place over a ten-year period. These were compared with 590 other houses of the same type in the same study area, selected at random from a street directory. The statistics are clear—and frightening. Houses with high field strengths occurred on both lists, but they were 40 percent more common on the suicide list than on the control list. You are very much more likely to be driven to the pathological extreme of killing yourself if you happen to live in a house too near a powerline.

It is possible that we respond passively to electromagnetic fields, letting our substance be juggled into alien patterns like iron filings on a tray. But it begins to seem more likely that we are so sensitive simply because we carry charges of our own. The beat of the human heart creates a distinct magnetic field. Its intensity is so low that to measure it properly you have to go way out into the country, far from electrical interference, but it is very definitely there, pulsing along. And recent discoveries in the new discipline of magnetoencephalography show that there is a similar field around every human head, extending a short way out into space and carrying a wave pattern that is directly influenced by mental activity.

There are those who even claim to be able to see it. The idea of an energy cloud or "aura" is very old. Blue flame has been associated with mystical exaltation since the enlightenment of the first holy man. It burned about the bush on Mount Sinai and hovers as a halo over the heads of saints and apostles. This luminous envelope seems to have nothing to do with the heat haze that surrounds us and can be "seen" by vipers with organs sensitive to infrared radiation just beyond the limits of the visible spectrum. It is more like Saint Elmo's fire, an eerie blue light that flames under certain conditions around masthead and yardarm of a wooden ship, clinging to anything metallic, there for anyone to see.

In 1939, the Russian electrician Semyon Kirlian discovered that it was possible to take "photographs" without lenses or light by putting an object and a color film in between two high-voltage, high-frequency electrodes that give rise to a visible discharge. The results are extravagant and beautiful. A human hand can look like the Milky Way, sparkling and twinkling against a glowing background of gold and blue. A leaf recently plucked from a living plant seems to shine with internal light that streams through its pores like candle glow through the holes in a Chinese lantern. Kirlian realized that he had stumbled on something that was at least a new art form, but it wasn't long before he began to suspect that it was very much more than that.

Kirlian found that leaves taken from the same species of plant showed similar patterns, and that these faded with time, but that a plant which was already diseased showed telltale differences. He took impressions of human fingertips and discovered that these changed too in ways that seemed to reflect the health and mood of the subjects involved. And he concluded that he had found a way of looking at "electrical signals of the inner state of the organism."

By 1970, the process had become highly sophisticated and was being used in several Russian laboratories interested in finding a mechanical way of pinpointing the energy flows said to be involved in acupuncture meridians. They found complex patterns of direct-current potential on the skin that correlated remarkably well with traditional acupuncture points, and were able to produce Kirlian pictures of these points that show them blazing like beacons in the night. But what really intrigued scientists in the West was a picture produced by one worker to show the "phantom" image of part of a leaf that had been cut away before the picture was taken. This, in one startling image, not only revealed the existence of a possible "energy body," but suggested that it was relatively independent and could continue to exist even after the destruction of the physical body. All at once, it seemed, we had scientific proof of the soul.

Kirlian blueprints began to circulate like copies of the Dead Sea Scrolls. Dozens of devices were built in laboratories and homes all over the world. Most of them worked well enough to produce new variations on the startling Kirlian displays. Magazines and books began to appear with superb color illustrations of people and plants in various conditions of sickness and health, either bursting with pyrotechnic energy or fading away into dark silhouettes like dying cities. International conferences were organized to discuss the Kirlian effect and there was talk of high-voltage emissions, radiation field photography, and the possibility of "bioplasma"—a high-tech version of mystic *prana*—the yogic source of an energy.

As more scientists became involved, the Kirlian mechanics came under closer scrutiny. We understand now that much of the effect depends on ionization of the air around whatever is being "photographed." And that the color of this corona depends to a very large extent on the nature of the chemicals and the amount of moisture that may be present. A hot sweaty person or a fresh moist leaf naturally differs from a cool composed subject or an old dry leaf. The shape of the display seems also to be affected by electric stresses that depend on the curvature of the surfaces being examined and on the shape of the air between them. These can both be radically changed, for example, by the simple expedient of pressing a fingertip harder down on the sensitive emulsion, altering the shape and size of its impression.

The difficulty of controlling all such variables has made it very easy for skeptics to dismiss the whole phenomenon. The official position at the moment is that Kirlian effects may be pretty, but they are simply artifacts produced by the process itself and tell us nothing about hitherto unknown radiations or energy fields. The "phantom leaf" effect, which has now been independently produced in Japan, Brazil, Britain, and the United States, is shrugged off as the result of inadequate cleaning of the electrode between photographs of the intact and the mutilated leaf. Orthodox science, it seems, has been saved yet again from the menace of bioelectricity.

But is it really that simple? I doubt it.

In Pennsylvania, L. W. Konikiewicz has set up the most carefully controlled Kirlian apparatus ever devised. He works in an environment chamber with constant temperature and humidity. He uses an independent stable oscillator to produce a consistent high-voltage supply. He does his research in cotton gloves and he and his apparatus are washed in distilled water and air-dried. He is given experimental subjects according to a double-blind system, which means that he knows nothing of their background. And yet, despite these elaborate precautions, he is able—simply by looking at the Kirlian pictures of their fingerprints—to determine sex and, in each of the female subjects, the stage of the menstrual cycle. He can pinpoint the day of ovulation with great accuracy and even tell whether a woman is on the pill or not. He has also succeeded in identifying patterns in the pictures that are characteristic of patients with cystic fibrosis (a hereditary glandular disease) and can identify those who show no overt symptoms at all, but are carriers of the recessive gene.

Diagnosis of this kind may depend on nothing more than chemicals in the subject's sweat, but there is new and impressive evidence to show that, along with chemical shifts, there are small but significant electromagnetic changes taking place in and around the body as a result of various diseases.

In Bucharest, the Romanian physician Ion Dumitrescu has taken high-voltage photography a step beyond the Kirlian process with the development of techniques he calls electrographic imaging or controlled electroluminescence. These involve the use of very high tension equipment and very short (one millionth of a second) and well controlled bursts of electrons that create transitory luminescent effects. In other words, instead of exposing a subject to prolonged radiation, Dumitrescu relies on a very quick "snapshot." And he can, with various forms of his apparatus, look either at a whole human body or at a single cell.

Dumitrescu's pictures of the body show the expected highlights on forehead, palms, and soles of the feet, places where sweat is being formed. But they also reveal patterns of electrical resistance on the chest as a result of the field produced by the heart—and indicate problems created by heart-muscle disease or blockages in blood supply. Areas affected by nerve malfunctions, such as ulnar paralysis or sciatica, show up as dead spots on the hand or foot. Duodenal ulcers flame right through the skin like miniature volcanoes. And malignant tumors appear as bright areas in the images, indicating magnetic voids. These pictures are not just another form of X ray, which simply sends a signal through the passive tissue and picks up shadows of the structures in its way. Electrographic imaging depends on the active patterns of electricity already in the body, letting these take their own picture with the help of the apparatus.

The spot of light that, Dumitrescu has learned, indicates an ulcer does not necessarily appear directly over the duodenum, that part of the human gut most liable to such perforation. It more often lies in a position that corresponds more directly to the traditional duodenum acupuncture point, and that seems to be connected to the organ itself by the path of least resistance through the body. This is important. The beauty of electrical imaging is that it draws attention away from spot symptoms and reminds us strongly of the connectedness of the whole organism. It gives us a more holistic view of a living being, which is probably closer to the truth.

The sum of our present knowledge points to the conclusion that all biological processes are fundamentally electric, and that central nervous systems in particular generate and transmit minute currents that regulate development, growth, and healing. We are electric creatures, living in an electromagnetic environment that may well have shaped our origins and continues to determine the direction of our evolution. The recent discovery of deposits of magnetic material in our bodies (we seem to have high concentrations in the nasal area) is being looked at for the moment only in connection with compass orientation—but it seems equally likely that we possess internal antennae that keep us tuned in to a great deal more than what happens to be going on only in our immediate sensory environment.

Said the poet Walt Whitman:

> I sing the body electric,
> The armies of those I love engirth me and I engirth them,

> They will not let me off till I go with them, respond to them,
> And discorrupt them, and charge them full with the charge of the soul.

There is no getting away from it. We are plugged in and switched on. All we have to do is respond.

9

The Wonder of Water

Water is a combination of two of the most common elements in the world. It is the simplest and most abundant of all chemical compounds, good old H_2O, but there isn't a scientist anywhere who can tell you exactly how it works.

Water breaks all the rules.

The laws of physics require that substances become more dense as their temperature falls and they shift from gas to liquid and finally to a solid state. Most do so, but water behaves in the expected manner only until its temperature reaches 4°C. Then something weird happens. Instead of shrinking further, water suddenly starts to expand, until at 0°C—the traditional freezing point—it has grown in volume by as much as 10 percent. Which is why rocks split, pavements buckle, and pipes burst on cold winter nights. And why there is still any life left anywhere on earth at all.

Were it not for this errant behavior, ice would sink. And it would not be long before every pond and pool, every arm of the ocean, was solid. All earth's water would be locked away and there would be no wells or springs, no streams or brooks, no humidity and no rain. Nothing but an ice age that went on forever.

Fortunately, things aren't like that; but only because water is odd enough to prevent such a disaster.

The laws of chemistry expect compounds to act as either acids or bases, settling on either side of a natural reactive divide. Most fulfill such expectations, but not water. It is one of the few known substances that can behave both as an acid and as a base, so that under certain conditions it is capable of reacting chemically with itself.

Water is colorless, odorless, and tasteless. It is gentle enough to support an array of delicate living things, but it is at the same time tough enough to corrode the hardest metal. Which is why almost every chemical reaction we know of takes place in an aqueous solution, and why there is no such thing as perfectly pure water. There isn't a container strong enough to hold it.

Were it not for this hyperactivity, no essential mineral could pass from the soil into the roots of plants, or flow up into its shoots and flowers. No animal could digest its food, or move nourishment about in its bloodstream. And there could be no absorption of life-giving gases by the moist membranes of leaves or lungs. Life, if it had ever got started, would long since have ground to an arid habit.

Happily, it hasn't; but only because water keeps astonishingly busy.

The secrets behind all this strange behavior lie locked in the structure of the water molecule. The combination of two atoms of hydrogen and one of oxygen is a powerful one, difficult to break but also inherently unstable. For water to exist at all, it has to be intricately laced. It is in fact held together by so many hydrogen bonds that it becomes an almost continuous structure. A glass of water is, in effect, one gigantic molecule with all its parts interconnected. And this gives it both extraordinary strength and an astonishing flexibility.

Ice is even more regular, forming the most perfectly bonded hydrogen structure known. Each molecule in it connects with four others, building up regular patterns like the six-pointed stars of snowflakes. These crystalline lattices are so precise that they seem to persist, hanging on even into the liquid state. Melted ice may look clear, but actually contains short-lived regions of crystals that form and dissolve many millions of times a second. It is as though liquid water remembers the form of the ice from which it most recently came by repeating the formula over and over again to itself, ready to change back again at a moment's notice. If one could take a photograph with a short enough exposure, it would probably show icelike areas even in a glass of hot water.

The strength of hydrogen bonds in water is most evident at its surface, where molecules cling together to form an invisible and highly flexible coat of liquid chain mail. The margins of any body of water, even one as small as a single raindrop, are extraordinarily elastic, giving it a strange and wonderful integrity, holding it in the form of smooth curves, waves, and spheres that have many of the properties of living things. Including an ability to act against the natural tendency of inanimate matter to seek the easy way out, to flow mindlessly toward equilibrium. A lot of the time water goes its own way, traveling uphill in direct contravention of entropy and the force of gravity.

Water molecules not only have a tendency to cling to each other, but they become attached almost as easily to other surfaces. They grasp at the edges of particles in porous soil and hold tight even to the smooth walls of veins and arteries and the mirror surfaces of glass tubes. The hydrogen molecules at the edge of any water surface reach out in all directions, grasping at any free oxygen to be found there and, once attached, use such anchor points to haul the rest of their substance along like boats warping up against wind and current.

And it is this special talent, called capillary action, that makes water not only physically and chemically unique but biologically essential. Without it, nothing would flow through frond or flesh; circulation would cease and life become impossible.

Given the fact that water is so strange and unruly, one might expect it to be rare. In universal terms, it probably is. Free water is hard to find in the cosmos, but on earth it abounds. There are 8 million cubic kilometers of it rising slowly up through the soil; 1.2 billion cubic kilometers sloshing about, covering almost 75 percent of the earth's surface; and 12,000 cubic kilometers more drifting about in the atmosphere as water vapor. Living things are abundantly irrigated, some of them almost waterlogged. Tomatoes and jellyfish are saturated at 95 percent, frogs and prawns reach 80 percent, and 75 percent of a chicken (more in some unscrupulous supermarkets) is liquid. It is a melancholy fact that even Raquel Welch is 65 percent water. On average we each contain about 38 liters of water and need to replace at least 2 of these every day. We can live without food for more than two months, but without water we die in less than a week. Water is unquestionably the world's busiest substance. Everything depends on it.

The nature of this dependence in our case is very precise.

Water absorbs heat reluctantly and, true to its nature, unevenly. In most other substances, the amount of heat needed to increase temperature by 1° is the same for each degree involved. But not with water. Between 35 and 40° C, water is unusually relaxed and most easily warmed. And this narrow range happens to coincide, if you believe in such accidents, with the usual body temperature of humans and most other active animals. This is not only convenient, but may be vital. It could even be the trigger that makes many living things so wonderfully responsive to their environment.

Giorgio Piccardi of the Institute of Physical Chemistry in Florence has always been intrigued by the vagaries of chemical reactions. Sometimes they work; sometimes they don't. He suspected that even simple reactions might be subject to influences beyond normal laboratory control and set up a lengthy experiment to test his hypothesis. Three times a day, every day for ten years, he poured bismuth oxychloride (a common colloid) into distilled water and recorded the time it took to form a cloudy precipitate.

The times varied enormously, but not randomly. There were short-term sudden changes that he was able to connect with changes in earth's magnetic field. And there were other long-term influences that correlated closely with the frequency of spots on the sun. Piccardi concluded that "water is sensitive to extremely delicate influences and is capable of adapting itself to the most varying circumstances to a degree attained by no other liquid." And that it was changes in the water that were influencing his reactions.

Water begins to look like something more than a convenient

solvent or a useful aid to digestion. Its flexibility and its sensitivity around normal body temperature make it an ideal go-between. A point of contact between ourselves and the cosmos. Something tantamount almost to a separate sense organ.

This line of thought has recently been explored by Theodor Schwenk, a German engineer who, in a creative book on flowing forms, suggests that water's sensitivity may be as great as that of the human ear. "A gentle breeze blowing over the surface of water immediately creases it into the tiniest capillary waves. . . . Water may be even more 'impressed' by a stone thrown into it, and it passes this impression on rhythmically to its whole mass. The great rhythms of the tides are a response to forces which work in the interplay of earth and cosmos . . . and for which, through its greater impressionability, the element of water is a receptive 'sense organ.' "

Schwenk goes on to describe water as "the impressionable medium *par excellence.*" He sees its boundary surfaces as receptors, made especially sensitive by the presence there of complex wave patterns that turn them into structures with some of the properties of living membranes. When an inert body of water is made to move, even by being shaken up in a closed vessel, its receptivity improves. Its sense organs open up and it changes physically as a result.

To document such changes, Schwenk prepared a number of identical bottles of water and had them shaken mechanically every fifteen minutes before, during, and after a total eclipse of the sun. When the eclipse was over, he introduced a fixed number of wheat grains into each bottle and allowed them to germinate without any further disturbance. The blades of wheat in water shaken at the time of the eclipse grew far less strongly than those in water agitated before or after the event.

Schwenk concluded that "a stream, bubbling mainly over stones, forms countless inner surfaces and tiny vortices, which are all sense organs open to the cosmos." And it passes on "impressions" so received to plants and animals and man.

The fact that all living processes take place in an aqueous medium makes it unnecessary to look for evidence of cosmic action on water outside the body. There are enough opportunities for an organism to be sensitized by what is taking place within. But it is a fascinating and provocative thought that a body of water deserves to be considered as an organism in its own right. As a creature that metabolizes, moving more quickly when it is warm and less dense, putting out new feelers, opening up extra sense organs as it flows from a cool forest glade out into the sunlit expanse of a summer meadow.

Small differences in temperature do seem to have quite dramatic effects on water.

In 1920, Prince Adolf von Schaumberg-Lippe of Austria found himself with a problem. He still owned 21,000 hectares of untouched forest at Bernerau and was anxious, in the depression that followed the First World War, to realize this asset. The forest, however, was 50 kilometers from the nearest road or river and the cost of transport made cutting it uneconomical. He tried to build a wooden chute to float the logs out, but there was too little water in the area to feed it—and there the project stuck until a young forester called Viktor Schauberger came up with a new idea.

Schauberger had noticed that streams seemed more dense where they were coolest, at the source of a mountain spring. He suggested that a chute might be able to support logs with less water if the water was cold, and proposed that the precious liquid be let out of the chute here and there, and replaced with fresh cold water from other springs and streams along the route. The experts poured scorn on this notion. "Water is water," they said. But Schauberger persisted, converting a chute at his own expense and proving his point in a triumphant demonstration in front of the prince and a crowd of skeptical forest-masters.

Part of the brash young forester's secret was cold water, but his success was due also in good part to another discovery. On a clear night in the forest, Schauberger had an almost mystical experience. He stood by a deep pool in a mountain stream and was astonished to see, down through the crystal water, that stones on the bottom were moving. He said later, "I did not trust my eyes anymore, when suddenly an almost head-sized stone began to move in a circular path. The stone was egg-shaped and in the next instant was on the surface, where it appeared to float, lit by the full moon. It was followed by a second and a third, until nearly all the stones of the same shape were on the top, while other more angular ones remained below and did not move."

He was not imagining things. His "dancing stones" were evidence of a phenomenon now known as cycloid or hyperbolic space curve motion, which has surprising effects on liquids. It is easily demonstrated. Fill a tall glass beaker with water. Drop an egg into it. The egg will sink to the bottom. But if the water is stirred in the right way, and this need only be gentle, the egg rises to the surface and stays there until the motion ceases. The specific gravity of the egg is only marginally greater than water and the motion has an effect equivalent to "concentrating" the water and giving it, at least in part, a density great enough to support the egg and let it float. It takes a stronger stir to float a rock of course, but in a pool of the right shape on a cold night with strongly flowing water already near its point of greatest density at 4° C, it can happen.

Schauberger's brainstorm was to apply this nocturnal insight to the problem of shifting heavy logs. He ignored the conventional wisdom, which insisted that logging flumes should travel by the shortest, quickest route downhill, and began to experiment with long meandering chutes that could take a little water and give it a lot of power by helping it to assume the swirling serpentine flow he had seen in the mountain pool. "Water in its natural state," he said, "shows us how it wishes to flow. We should follow its wishes."

He remembered the egglike shape of the floating stones and deliberately constructed wooden chutes with the same sectional proportion as the widest part of an egg. He nailed baffles and guides into the curves in the flume to impart a special spin to the water flow. He installed valves to siphon off warm water and introduce new cooler liquid at carefully spaced mixing stations along the line.

And he let his chutes follow easy, natural meanders all the way to their destination.

The results were sensational. After the initial success at Bernerau, Schauberger went on to build new flumes throughout Austria and the Tyrol, and on into Germany, Czechoslovakia, Yugoslavia, and Turkey, moving logs where no one else could manage it. The experts were livid. They couldn't understand his technique. They tried to copy his constructions, but every time they did, the logs remained stubbornly stuck in the chutes—moving only when Schauberger himself came in and made a few delicate and incomprehensible adjustments, altering the water temperature perhaps by just one tenth of a degree. The precise nature of his special chutes is still a mystery. The last of them was demolished in 1951 and all that remains is a few meters of worn film from a documentary commissioned by the Austrian Tourist Board in 1930. What it shows, however, is undeniable—logs churning down through strangely shaped wooden chutes at speeds and angles in direct contravention of accepted theory and expert opinion.

Schauberger went on from forestry to apply his hydrodynamic insights elsewhere, producing theories that remain highly controversial, but that are, in the light of his success in the log business, worth considering seriously. He believed that naturally moving water was a living substance, "earth's blood," that underwent a process of maturation, growing and changing and renewing itself. And that, if treated artificially, it "died." He insisted that scientists who studied water in the laboratory were looking at a "corpse" and were unlikely to learn anything useful about it. He was suspicious of "immature" water pumped up from deep underground, suggesting that, as it had not yet passed through the whole of its natural cycle, it was likely to be harmful to living things and might even be a cause of cancer. To provide healthy drinking water in areas fed by wells, he built a machine that provided "living water" by adding small quantities of vital metals and stirring it in darkness with a hyperbolic centripetal motion, allowing its temperature to fall in the process to "biological zero" at 4° C.

All of which sounds a bit cranky, though it has to be admitted that we have begun in recent years to become aware of the variety possible in water. Natural water has roughly 1 part of the hydrogen isotope deuterium in every 5,000 parts, but if this proportion is increased, we get heavy water, which can be used as a moderator, slowing down neutron activity in nuclear reactors. And we have polywater, a new anomaly produced by impurities that change the density and viscosity of water and even start it flowing spookily uphill. It is no longer quite so outrageous to suggest that the properties of water, both chemical and physical, may be changed by its experience. And anyone who has tasted natural spring water is left in very little doubt of its qualitative difference from city water that is used over and over again, passing from mouth to laboratory and back to mouth again without ever being allowed to touch the earth.

Schauberger believed too that the pipes in which water was being carried should be made of natural materials and shaped to produce the right kind of flow. He put copper guides into wooden pipes to give the water a sort of double twist that made it "sparkle with energy." And maintained that water treated in this way was inimical to pathogenic bacteria. Another hazy notion, perhaps—except for the fact that more and more natural processes, from the form of the DNA molecule to the lines of growth in muscle and bone, are now being shown to be essentially spiral. And that water put through particular rhythmic motions in a spiral pattern is demonstrably different. In the last decade, John Wilkes in the United States has produced and marketed a series of "flowforms," which send water pulsating down a figure-eight vortical meander that seems to give it a greater capacity to support plant and animal life.

In Bavaria, some of the high-forest farmers of Mühlviertel grow consistently better crops of potatoes and oats than their neighbors. Their soils and seeds are similar, but the successful farmers all practice an ancient rite known as *Tonsingen* or "tone singing." Toward evening each day in the growing season, they stir a little clay into a bucket of water, mixing it in well with a wooden spoon. And as they stir, they chant or sing directly into the container, letting their voices slide through a rich musical scale—on ascending notes when stirring counterclockwise, and with descending tones to go with a clockwise motion. The buckets are then stored in the dark, and early the following morning the farmers take this watery brew out into the fields and sprinkle it over their crops with a palm frond, like priests at mass anointing their communicants. Superstitious nonsense? Perhaps. But these clay singers enjoy a 30 percent greater yield than their less vocal friends.

A similar effect surfaced in a series of tests made in the Department of Psychiatry at McGill University in Montreal. They put twenty barley seeds into each of twenty-four pots of soil and deliberately wounded the lot by soaking them all in saline solution and drying out the soil in an oven. Half the pots, chosen at random in a double-blind experiment (which meant that nobody knew which were which), were treated with ordinary tap water. The other half received the same tap water, but only after it had been held for fifteen minutes in a sealed bottle between the hands of an Hungarian psychic who claimed to have healing powers. He certainly seemed to do something to the water, because the barley seeds receiving his help had a higher germination rate, grew taller, and weighed more than the control group—with the statistical odds against these differences occurring purely by chance at over a thousand to one.

The problems with muddy water that has been serenaded, and tap water that has been held in healing hands, and holy water from Lourdes or any of a thousand popular shrines, is that chemical and physical analysis shows no change in any of them. Whatever lies behind their apparent potency is inexplicable in terms of present-day physics or chemistry. We are left with indications that something important has happened, but no real knowledge of what that something might be.

We are aware of water's molecular integrity and are beginning to appreciate how this gives it a tensile strength as great as many metals. A very thin strand of the purest possible water is strong enough to lift itself over 3 kilometers into the air, forming an unbroken column as tall as the average depth of the ocean. But linked to this strength is an equally astonishing sensitivity.

In most natural systems, such as that which exists around a winter lake or pool, water commonly exists in all three states—solid, liquid, and gas—simultaneously. It covers all the options, flowing from one to another with the flux of energy, responding to even the most subtle changes in the environment; perhaps even recording them, filing the patterns away for future use in the same way that water remembers being ice and flows about muttering the secret formula to itself.

It may be a while before we are capable of consciously cracking the code in which such information is stored. But we are so liquid ourselves it seems likely that, unconsciously at least, we are already getting the gist of the message. It is something to wonder about.

There is no question that molecules of water, quite apart from their information content, are stores of extraordinary energy. They can absorb and release more heat than almost any other substance. The energy in just 1 liter of water is enough to keep a standard 60-watt bulb burning for almost one hundred hours. It is the cumulative effect of millions of such laden liters in flux that gives even an ordinary summer afternoon thunderstorm the energy of thirteen atomic bombs, and turns hurricanes into self-sustaining heat engines with the energy equivalent of half a million bombs the size of the ones that destroyed Hiroshima and Nagasaki.

Such mammoth exchanges of heat and energy take place all the time, powering the atmosphere, giving rise to the wind, distributing warmth and awareness about the globe. There are over 1.2 billion cubic kilometers, more than a million million million tons, of water on earth, covering almost three quarters of the surface, turning our planet into a soft blue sapphire in sharp contrast to the hard-edged look of apparently lifeless Mars. But such large numbers have a way of obscuring the truth. The truth is that if we reduced earth to the size of a classroom globe thirty centimeters in diameter, the average depth of the sea would be very much less than the thickness of this page. The deepest trench would be a barely discernible dent of less than a third of a millimeter, and the surface would be only just damp to the touch. And yet it is this thin film, this delicate membrane with its paradoxical properties, that keeps us alive.

That is the wonder of water.

It is time that we grew to know it a little better. We need to learn to watch, in the words of John Keats:

> The moving waters at their priest-like task
> Of pure ablution round earth's human shores.

10

The Immersion of Man

The fossils that decorate our family tree are so scarce that there are still more scientists than specimens for them to work on.

The roots of the human tree seem to be firmly planted in Africa, where every once in a long while one of them becomes exposed by wind and water and falls into the hands of a dedicated band of fossil hunters. Thanks largely to Richard Leakey and Donald Johanson—and to those working with them—the number of such finds has increased dramatically during the last decade. But it nevertheless remains true that all the physical evidence we have for human evolution can still be put, with room to spare, into a single coffin.

The study of human prehistory has become a complex, expensive discipline practiced by teams of specialists who are skilled in statistical techniques and adept at squeezing meaning out of apparently unrelated sets of measurements. This lends a new and welcome objectivity to paleontology, but not even the largest computer can conceal the fact that we continue to erect elaborate deductive structures around rare and isolated bones—and that we could be hopelessly wrong.

There is room in the empty spaces between the fossils for wide differences in interpretation. The reigning experts differ, often in noisy and newsworthy ways as each new discovery gives one of them a temporary advantage. But their public differences serve only to conceal a basic agreement between them—which, given the circumstances and the paucity of evidence on which their conclusions are based, is very strange indeed.

The current orthodoxy goes something like this: Twenty million years ago, in the mild period known as the Miocene, there lived a flourishing population of generalized, primitive, hairy apes. None of them looked like any that are around today, but a few had characteristics such as enlarged molar teeth that hint at a change of diet and a move toward greater humanity.

It is assumed that sometime during the drier Pliocene, between 6 million and 19 million years ago, one or more of these ancient apes moved away from the dwindling forest and out onto the hot, open savannah, where they became more and more bipedal—ultimately producing a creature with the erect body and bearing of a modern man, topped by the head and brain of an older ape.

This is precisely what Johanson found in the Afar Triangle in Ethiopia in 1974 and christened "Lucy."

Leakey doesn't dispute Lucy's validity as a descendant of the ancient apes, only her right to be considered as a direct ancestor of ours along a single line of evolution. He believes that we split off from a common ancestral stock somewhat earlier than Lucy's date of 3.5 million years ago, and that we have had larger brains than hers for a longer time.

So the two acknowledged experts in the field of human evolution differ only in the timing of the transition from Miocene ape to modern man. Both agree that the change took place during the Pliocene and that it was prompted by an increasingly rugged life on the high, dry plains of East Africa.

This is, on the face of it, a reasonable scenario. The only trouble with it is that there is absolutely no evidence to support it.

No Pliocene fossil bed containing any of the necessary missing links has yet been found in Africa. Beyond Lucy there is nothing but a gaping black hole in the record, stretching from 4 million to 10 million years ago. There is an arm bone from Kanapoi, part of a jawbone with a single molar in it from Lothagam, and another isolated tooth from Lukeino. These three lonely, worn little fragments from East Africa are all that have been found to bridge a gap of 6 million years during which the vital transition took place.

Modern apes seem to have sprung out of nowhere. Molecular evidence suggests that they are surprisingly close relatives of ours, but they have no established yesterday, no clear fossil record. And the true origins of modern humans—of upright, naked, talking, big-brained beings—is, if we are to be honest with and about ourselves, equally mysterious.

No one disputes the fact that modern humans and the living great apes had a common ancestor. We have enough characteristics in common for it to be clear that our lives diverged comparatively recently. We still share something like 98 percent of our genetic material with chimpanzees. The similarities between us and the apes are evident and easily understood. It is the differences that are perplexing. Why should our backs be straight, our skins bare, and our lives laced together with webs of words?

Somewhere in the genetic 2 percent that makes us uniquely human lie reasons to account for the fact that our posture, our locomotion, and our intellect should be so different from theirs. We seem to have spent a large part of the last 10 million years rushing through a series of evolutionary adaptations while the apes changed relatively little.

Why? What was it that made such changes necessary? Something must have happened to us that didn't happen to the chimps and gorillas. But what?

Theories abound and range, according to your taste, from environmental factors that drove our ancestors out of the forest, to banishment from the Garden of Eden by divine decree. In other words, we became erect, naked, and intelligent either because of a change of climate or due to an act of God.

Both theories are tenable. Scientists, of course, tend to favor the former, but it is important to understand that, in the absence of appropriate fossil evidence, it is actually no more susceptible to proof than any of the more traditional accounts of creation.

We ought, if we want to be reasonable about the whole thing, to be looking more closely at the yawning void in human evolution that lies between Lucy and the early Pliocene. And asking ourselves what might have happened during this period, between 4 million and 10 million years ago, that would have made such drastic changes in our bodies and minds both possible and useful.

There seems to me to be only one scenario that fits all the known facts. It makes no assumptions about genes, or tools. It presumes nothing about hunting as a crucial way of life. It is, in essence, a very simple theory that suggests that we are as we are simply because we once had an ancestor who spent a lot of time swimming. An ape that left the forest and, instead of going directly out onto the open savannah, spent some millions of years messing about in the water. An aquatic ape.

It has happened before in other groups of animals. Quite often.

Long before the first mammals came into being, an air-breathing land-living dinosaur went back into the sea and stayed there long enough to grow flippers. It is even possible that some of these "fish-lizards" from the Cretaceous still survive in the more remote corners of central Africa. An American expedition from the University of Chicago in 1981 and a local group from the Brazzaville Zoo in 1983 both reported seeing something very like a giant aquatic reptile in the Likouala region of the People's Republic of the Congo.

Seventy million years ago, a group of warm-blooded early ungulates, grazing on the banks of tropical rivers, took (as hippopotamuses still do) to the water and evolved to fill an ecological niche left open by the sudden extinction of most of the ruling reptiles. The whales and dolphins are direct descendants of these opportunists. Twenty million years later, it was the turn of the elephant family. A transition that has given us the beasts we now know as sea cows—the manatees and dugongs. And between 25 million and 30 million years ago it was relatives of the bears and the dogs that submerged themselves to become seals, otters, and beavers.

If reptiles, ungulates, and carnivores—not to mention birds, insectivores, marsupials, and rodents—can all take to the water, why not a primate? In the mangrove swamps of Indonesia right now, there is a primate—the proboscis monkey—that spends a great deal of its time in the water. These floppy-nosed monkeys have not yet developed any obviously aquatic characteristics, but perhaps it is simply a matter of time. An ape that went swimming more than 10 million years ago and was still in the water millions of years later might well have changed in some important ways.

It is likely, for instance, that (like whales) aquatic apes would have lost their hair. With the exception of a burrowing rat and one highly artificial breed of Mexican dog, all hairless mammals in the world today are either aquatic or spend a great deal of their time wallowing in water and mud. Fur is useful as an insulator against heat or cold only when it is dry. Wet, it loses the ability to trap a protective layer of air next to the skin and becomes a liability. The fact that we, the only naked apes, have retained long hair on our heads may have something to do with protection from the sun. But it could also be important for another reason.

During the last twenty years, we have discovered to our astonishment that human babies are able to swim well almost from the moment of birth. In the 1960's, Dr. Igor Tjarkovsky, a gynecologist in Moscow, pioneered a system of underwater childbirth—largely for the therapeutic benefits he found it brought to mothers. But there were also unexpected side effects. The high proportion of fatty tissue in newborn babies makes them tremendously buoyant. They pop straight up to the surface and show strong reflex paddling movements of the arms. They also demonstrate a surprising ability to control their breathing and seem to be perfectly happy with their heads underwater. They stop trying to breathe when submerged and behave wonderfully calmly, gazing around with wide-open eyes and no sign of struggling or fear. They appear to enjoy the freedom of weightlessness. Dr. Tjarkovsky's own three-month-old daughter became so proficient that she was soon able to hold her breath for more than three minutes, which is as long as the most practiced *ama* women divers in Japan.

For children like these, an aquatic existence poses few problems. But it would have helped an active mother, busy feeding herself and her offspring on mollusks and crustaceans, to keep contact with her babies in the water if she had something they could cling to. As it happens, she did have something—the hair on her head. Which helps explain why we have such long hair there, why we have kept it there when all the rest of the body fur has been reduced, and why, among us, it is only males who tend to go bald and become completely naked apes.

Allied to our instinctive ability to swim is a fascinating reflex that comes into action as soon as water touches the skin on our faces. We respond by automatically reducing our heartbeat and the rate at which the body uses oxygen. This "dive reflex," which cuts the normal pulse rate from 70 to 30 beats a minute, is something we share with whales and seals, but is not known to exist in any other land animal.

Another consequence of immersion has been a change in sexual behavior. All aquatic mammals mate belly to belly. It is difficult for them to do otherwise. Their heads and spines and hind limbs have become arranged so that they lie in a straight, rather stiff, line. This is the most efficient arrangement for swimming. It is better streamlined, but it makes the rear-entry method practiced by other mammals during sex impossible. So the female sex canal in whales, seals, and dolphins is tilted forward and they make love face to face, as we do.

But we are the only primates who do. No other monkey or ape can. And it is difficult to avoid the conclusion that our uniqueness in this respect must have something to do with the fact that, like the aquatic mammals but unlike any other primate, we too went through a stage of prolonged immersion.

Humans are the only mammals that walk about constantly on their hind legs. A few marsupials hop around on powerful back springs, but all of them also rest on tripods completed by the addition of a muscular tail. We are unique in being totally bipedal—and have to wonder why.

Two legs are not necessarily faster or more energy-efficient than four. Being bipedal makes it possible to see farther over tall grass and it does free the hands for carrying things, but in almost every other way it is a highly precarious business. We fall over and injure ourselves a lot and suffer all our lives from a painful variety of strains, hernias, and backaches. Ground-living monkeys rise up on their hind legs only when they really need to. There must, it seems, have been a very pressing reason for us to have so completely abandoned the safety, speed, and stability of having a leg at each of our four corners. Something really vital. Something as basic, perhaps, as breathing?

Try crawling into the ocean on all fours and see what happens to you. As the water gets deeper, you are virtually forced into an upright position. And when the water is too deep to stand at all, even on tiptoe, it is still easier and less tiring to float, as seals and otters do when they rest, in a vertical posture. And having been vertical for a good part of each day over millions of years, an aquatic ape would certainly have found it easier and more natural to be bipedal when returning to life on land.

This is the essence of the aquatic ape theory. It looks at the mystery of a creature that persists in walking about on its hind legs without hair on, and suggests that these peculiar habits were acquired underwater. And that, having spent a while submerged, we came back or were forced back to the rigors of life on dry land wonderfully well prepared to leap ahead of our landbound relatives, who went straight from the forest to the savannah without benefit of an immersion in between.

The first person to think seriously about the sea as a force in human evolution was Sir Alister Hardy, a British zoologist who served his apprenticeship as a marine biologist on oceanic expeditions earlier in this century. His awareness of a rising tide of consciousness among living things led him to suggest that there were psychic forces acting in evolution. And in 1960 he drew a direct connection between man's mind and the sea. "I see our ancestor," he said, "becoming more and more of an aquatic animal, going further out from the shore; I see him diving for shellfish, prising our worms, burrowing crabs and bivalves from the sands on the bottom of shallow seas, and breaking open sea urchins, and then, with increasing skill, capturing fish with his hands." And, he suggested, it was this aquatic phase that set our ancestors apart from the apes and gave us both the opportunity to grow a bigger brain and the incentive to use it.

In 1967, Desmond Morris referred in passing to Hardy's thesis by suggesting that his "naked ape" might once have gone through a "christening ceremony." And in 1972, the Welsh dramatist Elaine Morgan drew on the aquatic theory to lend support to her argument that our thinking about evolution has been exclusively male oriented and has led, as a result, to major misinterpretations of sexual response and the origin of social behavior. It was, she suggested, the female ape—the foraging partner—who led the way into the shallows of the sea and was the first to start using tools to break through the hard shells of food items she found there.

Ten years later, in 1982, Morgan took her thesis on the Descent of Woman to its logical conclusion by writing *The Aquatic Ape.* In this elegant book, she goes through all the arguments for a recent watery stage in our evolution, ending on a new and telling note with her observations on why it is that humans talk and apes don't.

Aquatic animals live in an environment where it is often difficult to see or smell each other and, as a result, all have developed powerful systems of vocal communication. Whales even "see" with their ears. They and dolphins have become very specialized, very different from us, but there is little anatomical difference between ape and human vocal chords. The single reason we speak and they don't is that we have acquired conscious and voluntary control over the air channels that power the larynx in the throat. And, as Elaine Morgan points out, it is hard not to come to the conclusion that this ability was acquired as a direct consequence of learning to hold our breath as we dived deeper and deeper underwater. No ape has ever had to do so and none, not even the most rigorously trained, has ever been able to get further than the rather indistinct pronunciation of the four simple words *papa, mama, cup,* and *up.*

I spent some of my most formative years studying under one of the great masters of our time—the anatomist Raymond Dart, the man who discovered the first of the African ape-men in 1925. Dart is blessed with a special sensitivity, a kind of scientific second sight, and it was he who—as we worked through the finds from Taungs and Makapansgat—taught me to look beyond the bones for meaning. It was he who trained me rigorously in scientific procedure, but never allowed me to lose sight of the importance of intuition. "It has to *feel* right," he said.

There is something about the aquatic age theory that feels right. It is certainly a radical reinterpretation of the known facts and it is easy to understand why the anthropological establishment continues to hold it at arm's length. But it is a very appealing hypothesis to a biologist because it fits so well with what we know of other animals. It puts man and his history into a comfortable and meaningful evolutionary perspective. And, as Dart would say, "I have a sneaking feeling . . ." that it is probably right.

Proving that it *is* right is another matter.

There are some useful geological clues. We know from the evidence of old marine terraces that the ocean level in the Pliocene changed dramatically. At one point, the sea came flooding in and inundated large areas of what are now East Africa's coastal plains. Throughout the area, there are ladder formations of raised beaches—some of them now perched 200 meters above present sea level. This means that the rather straight line of the existing coast was then broken by huge bays of shallow water that swept a long way inland, turning isolated hills into offshore islands. Populations of ancient apes marooned on such islands would have found that they soon ran out of their usual forest foods and that they were forced to change their diet, relying more and more on what they could find along the new shorelines.

These floods lasted over much of the Pliocene, ending only with the later arrival of hotter, drier times about 5 million years ago, when the sea left the aquatic apes behind and someone very like Lucy walked back across the floodplains to the shelter of the Great Rift Valley. It is significant that most of the fossil finds in East Africa are still being made along the shores of old lakes that decorate the area, acting, as they always have, as magnets for animal and human life. They would have been especially attractive to a displaced aquatic ape in search of the foods it had learned to like best.

The great gap in the fossil record—the continuing inability of paleontologists to find any protohuman remains older than 4 million years—may simply be due to the fact that they are looking in the wrong place.

Leon La Lumiere of the Naval Research Laboratory in Washington, D.C., has singled out the Danakil Alps on the coast of northern Ethiopia as a more likely site. He argues, persuasively, that this plateau was once an island not too far offshore, that it has geological deposits of the right age, and that it lies close to where Lucy was found. He may be right. It is certainly worth taking a look there, but I have a feeling that it is altogether too far north. I suggest that most fossil finds have been made there in Northeast Africa for reasons that have more to do with recent history than with ancient geography.

Louis Leakey happened to live in Nairobi and quite naturally started looking for things in his own backyard. And it was his early finds at Olorgesaillie and Olduvai (both just a day's drive from his home) that have kept people looking and working in this part of the Great Rift Valley ever since. It was Richard Leakey's even more exciting finds a little farther north, near Lake Turkana, that sent Donald Johanson and his team looking farther north still—just beyond the boundaries of Kenya and out of the Leakey area of influence.

In South Africa there is a similar, equally restricted, search area that grew up around Raymond Dart and his students in their backyard. But between the two, between South and East Africa, is a spatial blank, every bit as large and just as disconcerting as the temporal gap in the fossil record.

This is where I believe the crucial finds will be made.

The coastline in the early Pliocene was longer than it is now. We can, however, plot it quite easily from the geological evidence and eliminate long tracts of it as suitable sites for the simple reason that they would have lacked, and still lack, adequate fresh water. Our swimming ancestors may have spent most of their days in the salty sea, but they would have needed to drink large quantities of fresh water—and are most likely to have gathered around the mouths of the bigger and more reliable rivers.

There are many possible rivers, but I suggest that we can narrow the field down yet again by assuming that we need only look along those rivers that were large enough to have provided highways back into the interior—up into the Rift Valley where their descendants have been found.

And this process of elimination leads me, every time I make my calculations, to one area in particular—to the valley of the Zambezi River, which lies right across a natural gateway to the southern end of the Rift, where it opens out onto the old coast in southern Malawi.

This, I suggest, is where we should be looking. This is where I have already begun to look, walking along the hills in search of appropriate fossils on the edge of an ancient shore. It is a task complicated at the moment by drought and by a guerrilla war being waged against the government of Mozambique. But there is something about the area that feels just right. My intuition tells me that it is here, perhaps in a gully exposed by the very next rains, that someone is going to stumble across the bones of a being that will give us food for new thought.

We are not going to find webbed footprints on an ancient beach, but I predict that the context of the find, the evidently marine ecology, will confirm Alister Hardy's feeling that somewhere between the ancient apes and modern man lay a revolutionary being that can only be known as *Homo aquaticus.*

11

The Biology of Bias

"For every action," said Sir Isaac Newton, "there is an equal and opposite reaction."

Generally speaking, he was right. Duality seems to be a law of our universe, emphasized by light and shade, positive and negative, male and female polarities—all nicely balancing each other out.

Even our brains are symmetrical, sliced down the center into two identical hemispheres, each a mirror image of the other. The left side of the brain controls the movement of the right side of the body. And vice versa. All very neatly packaged.

But this apparent symmetry is purely superficial, because the vast majority of human beings are deeply biased. Between 80 and 94 percent of us are predominantly right-handed, which means that the left side of the brain has taken over most of the responsibility for motor control of the body.

This is a unique situation in the animal kingdom. It is true that individuals of other species do sometimes prefer one paw to the other. I once worked with green acouchis—a group of long-legged South American rodents that trot about the rain forests like guinea pigs on stilts, burying every nut that falls from the canopy. I discovered that some recognizable individuals consistently used one foot more often than another in digging, and had a matching preference for hiding hoards of food either to the left or to the right of conspicuous landmarks. But viewed as a whole, I found that acouchis in Brazil were happily ambidextrous. And the same seems to be true of all other animals ever examined.

We are the sole exceptions. Most of us now go around writing, throwing, lifting, and eating things with out right hands—and looking askance at those who don't. But it was not always so.

Stone Age tools show no apparent bias. Paul Sarasin, a French anthropologist, has made a meticulous study of the implements found at Le Moustier in the Dordogne—where inhabitants of the famous caves made over fifty different kinds of flint knives, scrapers, and points. Some of these Mousterian tools show reverse beveling, being sharpened on the left, but an almost equal number lean to the right. So Neanderthal man, it seems, had no particular preference in the matter of handedness.

An analysis of early cave paintings reveals that many of these were drawn with the left hand. One of the sure signs of such activity is a face shown in profile, facing right. A right-handed artist will usually produce a portrait that looks left—try a quick doodle for yourself and see. But surveys of Paleolithic rock art from around the world produce roughly equal numbers of human figures looking in each direction.

Living Stone Age people, such as the aborigines of Australia or the Kalahari Bushmen, have remained largely ambidextrous, but something changed for most of us about ten thousand years ago. And it seems to have happened first in early Bronze Age communities of the Fertile Crescent in the Middle East.

It was here that agriculture began when wild cereals were first cultivated and harvested, providing sufficient surplus food to support the first large permanent settlements. It is here that we find evidence of the earliest farming tools—simple flint-bladed sickles—and it is obvious from the shape and wear of these that not a single one was used in the left hand. They are still not. To this day, there is no such thing anywhere as a left-handed sickle or scythe. It looks almost as though farming provides a stimulus that is largely localized in the left side of the brain, involves the right hand, and disturbs natural equilibrium in ways that have in the end produced a different kind of human being.

Agriculture, when it first began, certainly required new skills. The repeated exercise of a special technique like scything called for deftness, endurance, and consistency of a sort that no hunter ever had to produce. Tracking, stalking, and killing were never easy, but compared to cultivation they are playful activities. Farming is hard work. It demands economy and accuracy and the adoption of a habitual approach. The tools involved can be complex and take a great deal of time and trouble to make. They are specialized enough to require that each be designed for use by one hand *or* the other, leaving no possibility of later change. The user has little choice in the matter; it is the maker who decides. And once such complex tools had been made, they became precious. Many were handed on from fathers to sons, who then had to learn to use them in precisely the same way, making the bias inherent in the tool effectively hereditary. So, all in all, there were very good reasons, right from the beginning, for a one-sided orientation during the Neolithic Revolution. But why did the right side win?

Since the Traffic Act of 1772, vehicles on British roads have been required to keep to the left. A nursery rhyme of the day puts it very succinctly:

> If you go to the left, you are sure to be right,
> If you go to the right, you are wrong.

The same rule applies in Japan and in a strange consortium of other countries—like Indonesia, Iceland, Somalia, and Yemen—which seem to have little else in common. The rest of the world, perhaps under the influence of the motor industry of Detroit, keeps to the right. Why the difference? Well, there are any number of theories.

The leftists point to the ancient need for a coachman to keep his whip hand, his right hand, free for argument with approaching vehicles and highwaymen. And those on the right, in France at least, pay homage to Napoleon—who won several spectacular victories by reversing the traditional order of battle and attacking from the right flank first. In defense of the latter idea, it has to be said that everyone in Europe still drives on the right, except (until very recently) the Swedes, the Bohemians, and the British—the only three major communities in the area *not* to have been conquered by Napoleon.

The origin of rules of the road probably lies in historical quirks of this kind. They are simply conventions, codes of common practice, which owe little to natural history. But the same cannot be said of sickles, which are everywhere the same; or of other evidence that suggests that there may have been good biological reasons for early bias, once it became necessary to be to the right rather than to the left.

Beginning in 1909, Sir Cyril Burt—then professor of psychology at University College in London—became hopelessly embroiled in a dispute about the inheritance of intelligence. Much of his work is now discredited by suggestions of fraud, but as part of his interest in the origins of behavior, Burt did some fascinating pioneer studies on left- and right-handedness. And these are still worth reading. He pointed out that our bodies are naturally asymmetric in that the heart and most of the stomach lie to the left of the midline. It follows, he suggested, that a warrior or worker will naturally turn his left and more vulnerable side away from potential danger. And trust more to the solid support of the right side of the trunk, which weighs a little more, using the right hand more often for wielding heavy instruments and weapons.

Burt wondered whether there were sex differences and whether women would show a similar preference. So he went out into the streets and looked at the way in which women do things like carrying babies. He found, to his surprise, that 73 percent do so on their left arms only. At first sight, this is a strange bias, which means that the child has to wrap its right arm around Mother's neck and, spending a lot of time in this position, ought to become predominantly left- rather than right-handed.

Burt confirmed this unexpected bias, and showed that it was not peculiar to the British, by going to medieval and Renaissance paintings of the Madonna and Child, finding there that Mary, more often than not, is also shown with the infant Jesus on her left hip. More recent studies in a number of countries have gone even closer to the heart of the matter. They too confirm the fact that women everywhere do tend to hold or carry their children on the left, perhaps as a way of keeping their own right hands free for other things. But suspicion grows that the real reason for this preference is that mothers have discovered, by trial and error, that their babies are better behaved, more easily comforted, if they are held to the left breast—where they can hear Mother's heart more clearly.

During the last months in the womb, a developing baby's world is dominated by the metronomic thump of its mother's heart. It is a distinctive sound that goes on night and day, pounding away in a rhythm that is peculiar to her alone. It pervades a secure world of constant temperature and instant nourishment. And later, after expulsion into an environment that must seem incredibly hostile by comparison, it is the sole reminder of those early, relatively carefree times. Studies have shown that newborn babies *do* respond very dramatically to recordings of heartbeats and are even able to pick out their own mother's distinctive rhythm from a number of others.

Most of us therefore end up clasped in Mother's left arm, held up close to her heart, with left ear and cheek hugged tight against her breast. We spend a large part of our most formative early months looking constantly to the right. And even when laid flat on our backs tend to show what is known as a tonic neck reflex attitude, lying with head facing right, looking at the hand that will, in most cases, come to be the one that we favor later when we need to coordinate movements involving the hand and the eye. At about six months of age, most children begin to show a preference for the right hand when reaching for things. And by the age of eighteen months will instinctively put out that hand to catch a ball thrown unexpectedly.

So, although we are basically ambidextrous, there was a time in our evolution when it became necessary for us to choose one hand over the other to perform a particular task. And it seems that, due to both anatomical and environmental factors, the choice was largely preordained. As farmers in the Bronze Age, or in almost any activity now, we cannot help being largely right-handed. It might seem a small thing, but the results of this slight shift in emphasis have been quite extraordinary, sometimes even catastrophic.

In early Roman times, citizens wore togas, simple flowing robes that had just one pocket or *sinus* on the left-hand side. This soon became known as the *sinister* side—a word that, in English, now means not only "left," but also "evil," "wicked," and "criminal." And so it goes in most other languages. Left in French is *gauche,* which also means "awkward" and "clumsy." In Arabic it is *shamal,* which is another way of saying "unclean." In Spain they describe someone who is very clever as *no ser zurdo,* a phrase that translates directly as "not being left-handed." A left-handed compliment is widely understood as an insult. And a left-handed marriage is one between a man of high rank and a woman of low rank, in which the children have no right to their father's possessions or title.

Everywhere in this right-handed world there is a prejudice against being left-handed. Part of this antipathy is the characteristic response of any majority toward a minority. It feeds on suspicion of anything or anyone that dares to be different, anything that goes against nature. But it also goes a lot further than that. There have been times and places in which being left-handed was lethal.

The Devil, of course, is left-handed. What other assumption can right-handed and right-minded humans make? So in Europe during the Middle Ages it was enough to invite suspicion of black arts and witchcraft if you ate or wrote with your left hand—or even so much as dared to walk counterclockwise around a house or shrine. Most portraits of the hapless Joan of Arc show her wielding a sword defiantly in her strong left hand.

The left is still widely regarded as being at least unlucky. After spilling salt, many people in England throw a little with their right hands over their left shoulders, in the Devil's direction. They never drink a toast or pass wine with the left hand and some make a point of entering or leaving a house with the right foot first. In Mexico, medicine is never taken with the left hand. And in Greece it is specifically the left hand, palm open, that is used as a gesture to represent the evil eye. The red palm prints that appear on the walls of houses in some Arab countries, to ward off the effects of such unfortunate glances, are always those of the good right hand.

Christian ritual is overwhelmingly dextral. The wafer and the wine at communion are given and received with the right hand. The sign of the cross and the benediction are always made with the right hand. All paintings show Jesus giving his blessing with two fingers of the right hand. To do so with the left is regarded as blasphemous and still forms part of the satanic Black Mass. There has never been a left-handed pope.

Chinese belief revolves around the cosmic principles of yin and yang, which are seen as opposites in equilibrium. But it is yang—the active, male, positive power—that is set on the right-hand side. The left hand is yin—symbol of things passive, female, and negative. In parables, an enlightened being always takes the right-hand way at a fork in the road, because this is the one that follows the eightfold path recommended by the Buddha. Gautama himself is said to have descended, as a white elephant, to enter through his mother's right side.

Islamic practice honors the right hand far above the left. Before prayer, the right hand is washed first, then the left. The Koran is handled only with the right hand. Meals are taken with the right arm bared to the elbow, using the thumb and two fingers of this hand alone instead of instruments. The left hand is reserved for toilet functions, which, though necessary, are considered unhygienic. The left is the unclean hand and everything associated with it is tainted. No Muslim prince can be comfortable traveling in a motorcar with the steering wheel on the right, because this requires him to sit on his chauffeur's left-hand side (or behind the chauffeur, which is not much better).

Tantric writings of India associate the right hand with being male, conscious, good, pure, sacred, joyful, healthy, progressive, active, and filled with light. While the left, inevitably, is female, unconscious, evil, impure, profane, sorrowful, unhealthy, regressive, passive, and enveloped in dark.

With this weight of official opinion against the left, it is hardly surprising that there are few heroes in history who have been conspicuously left-handed. The sole exceptions seem to be Joan of Arc, who came to an unhappy end, and Ehud, the biblical Benjaminite who took the court of Moab by surprise, stabbing their king with his left hand while the royal guards were watching for assassination attempts from the right. There is, on the other hand, no shortage of left-handed villains—including Jack the Ripper and Billy the Kid. The Ripper was never actually caught red-handed, but it is clear from the wounds he inflicted that this most notorious of all mass murderers was a natural left-hander.

There is no evidence whatsoever that being left-handed predisposes anyone to a life of crime, but there are now a number of studies that link left-handedness with certain disorders.

Being left-handed is sometimes associated with damage to the left hemisphere of the brain. The left side of the brain is dominant in right-handers and it has been suggested that left-handers may simply be right-handers who have had to adapt to overcome this handicap. But it has never been demonstrated that more than a tiny fraction of left-handers actually have brain damage of this kind.

A recent study at the Beth Israel Hospital in Boston shows that left-handers are, on average, around twelve times more likely to suffer from learning disorders such as dyslexia than right-handers. In most people it is the left side of the brain that specializes in the production and comprehension of speech, and it begins to seem possible that an abnormality in this hemisphere may be jointly responsible both for dyslexia (word blindness) and for a shift to right-hemispheric (left-handed) activity.

Both left-handedness and learning disorders are more common in boys than girls. Specialization on the right side of the brain can be demonstrated as early as six years of age in boys, while no marked differences between the left and right sides of girls' brains appear until they are about thirteen. Norman Geschwind in Boston suggested that in boys, the normal growth of the left brain (which has to do with language) is delayed by the presence in the body of the male hormone testosterone. This would explain why boys are usually slower than girls in their linguistic development, and why, as adults, they have to devote a far larger area of the left brain to language activity. In support of this theory is another recent discovery, that spatial abilities (which are localized in the right brain) are impaired in men who suffer from male hormone deficiency.

There is also a high incidence of left-handers of autoimmune disorders. These involve a failure of the body's normal immune system to recognize and take action against invading organisms. And such failure leads frequently to inflammatory diseases of the bowel and gut, such as ulcers and colitis. The fact that these come to be linked with left-handedness may be due simply to the fact that both conditions arise initially as a result of hormone deficiencies—and may be controlled by providing additional testosterone.

These are interesting discoveries, but they carry with them the very real danger of reclassifying left-handedness as an infirmity, as a disease that needs to be cured. It has been seen that way for centuries, during which every attempt was made to force children who showed "sinister" tendencies to do things the "right" way. The results, very often, were children whose left brains were short-circuited in a way that interfered with normal speech and left them stuttering.

Demosthenes, the Greek orator; Cicero, the Roman philosopher; Charles Lamb, the English essayist; and Charlemagne, the French emperor, were all chronic stammerers who ended up with severe speech impediments as a result of being forced, against their natural inclination, to become right-handed. There is no record of the number of children who have been similarly tortured, but lacked the courage and determination that helped these great men rise above their handicap.

Despite widespread prejudice, perhaps even to a certain extent because of it, there are positive characteristics that are common to many left-handers. They share an independence, a tenacity, an intelligence and self-sufficiency that have produced some remarkable talents. Prominent left-handers include the artists Leonardo da Vinci, Hans Holbein the Younger, Paul Klee, and Ronald Searle; musicians such as C.P.E. Bach, Cole Porter, Harpo Marx, and Paul McCartney; the actors Charlie Chaplin, Danny Kaye, Judy Garland, Betty Grable, and Terence Stamp; and numerous sportsmen, among them Babe Ruth, Rod Laver, Jimmy Connors, and John McEnroe.

There seem to be very few left-handed writers of repute. The only one that comes readily to mind is Lewis Carroll. The fact that our language centers and our right hands are both controlled from the same, the left, side of the brain may be significant. Our centers of spatial awareness and rhythm are in the right brain and there is no comparable shortage of great artists or musicians who are left-handed; or who, like Maurice Ravel, Richard Strauss, and Benjamin Britten, have gone out of their way to compose for left-handed performers.

The proficiency of left-handers in sport is easier to understand. In individual events like golf, there is no problem as long as the equipment is appropriately adapted. And in conflict sports, being a left-hander can even be a positive advantage. Such players put themselves, and the ball, into positions that many right-handers find awkward and difficult to deal with. In baseball, the batter traditionally faces east to keep the late afternoon sun out of his eyes. So the pitcher looks west, and one who throws the ball with his left arm will swing round from the south side—hence the term southpaw, which is now more commonly applied in another sport, to boxers who lead with their left. Outside the ring and off the field, however, these advantages rapidly disappear. In everyday life, left-handers everywhere still face the problem of scissors, irons, potato peelers, nutcrackers, T squares, and checkbooks that are all designed with the right-hander in mind. And all these implements help to perpetuate the myth of normality attached to doing things the "right" way round.

As an anthropologist, I can understand how the bias persists. But as a biologist, I have some difficulty in seeing how it came into being. Sitting on Mother's left hip seems to have helped the tilt toward the right, but I have to admit that I find it less than completely convincing. I cannot help feeling that there must be another, more pressing reason for the dominance of right-handers in our species. There must have been a time in evolution, quite apart from the need to wield a nifty sickle during the Neolithic Revolution, when being right-handed conferred on us a distinct and selective advantage. It must once have been critical.

The best idea I have heard so far comes from Peter Irwin, a pharmacologist with a drug company in Switzerland. He works on the effects of drugs on the brain and has discovered that left-handers are more sensitive on the whole than right-handers to chemicals that act on the central nervous system. For some reason, left-handers show more marked reactions to these substances on both sides of their brains. This may explain why they are unusually susceptible to learning disorders, immune diseases, and even epilepsy—all of which are conditions under central nervous control. But, as Irwin points out, it also provides a possible explanation for the majority status of right-handers. If right-handed individuals are less sensitive to chemicals that affect the brain, they would have been at an advantage in those early days, long before farming, when we were feeling our way through the world as food gatherers, and would have been exposed, every day, to the risk of being poisoned by toxins in new and unfamiliar plants. In this situation, being left-handed was in itself no direct disadvantage. But the linkage of left-handedness to a greater sensitivity to such dangerous plants could quite easily have killed off enough of our ancestors to leave right-handers in the dominant position they now enjoy.

I rather like that idea, but I am even more attracted to attempts that have been made to foster manual equality—to create people who are completely ambidextrous (or ambisinistrous). These began with the Athenian philosopher Plato and have been revived several times since, most recently by Lord Baden-Powell—the founder of the Boy Scout movement, whose members deliberately greet one another with a left handshake.

Ambidextral campaigners have been accused, like Esperantists, of being a bit cranky. But I believe that they are both on the right track. If you look at the truly great minds in human history, you will find that they were always people who were able to use both sides of their brains to equal effect.

Albert Einstein is a classic example. He was a great scientist, a supreme rational thinker, someone highly skilled in the mathematical use of his logical left hemisphere. But all his best ideas came to him as pictures and images. He hit upon the theory of relativity not as a logical deduction, but while gazing up at the sun through half-closed eyelids, letting his right brain run free while lying on a grassy hillside one summer afternoon. There are probably thousands of people around who have had similar insights. I get effusive letters from a number of them every year. But they make little sense because, unlike Einstein, they lack the ability and the left-brained skill necessary to communicate their revelations effectively to others.

We in the West are at a singular disadvantage in this respect. Our training, our education, and our environment are all unmercifully biased to the right. We learn to read from left to right and lay great emphasis on rationality, on the ability to express ourselves verbally, and on the importance of analytic thinking. We have become slaves to the left hemisphere and the right hand.

For those who are lucky enough to grow up in cultures without such a strong bias, things ought to be a little easier. Learning to read from right to left, and learning to do so with a script (like Japanese *kanji*) that requires complex pattern recognition, are ways of ensuring that the right brain and the left hand get their fair share of vital exercise.

The object is, in the end, to become fully aware. To use both hemispheres of the brain in balance, rather than temporarily suppressing one in order to make use of the other. There is evidence that, in the course of evolution, there has been a steady thickening of the corpus callosum—the bundle of fibers that connects the two sides of the human brain with each other. As a species, we seem to be moving in the direction of greater communication with ourselves. This is all to the good and is cause for optimism that we might eventually succeed in rising above the basic biology that seems to lie behind our bias.

Prejudice may be natural, but it isn't necessary anymore.

12

The Essence of Balance

Some of my best friends are cannibals.

They live on the Casuarina Coast, the delta area of Irian Jaya—Indonesian New Guinea. It is a flat mangrove swampland, a place of water, mud, and lush vegetation. An area ruled by rain and tide. At low tide, mudbanks extend for miles out into the sea, and at high tide, the sea runs right into the forest. It is impossible to say where water ends and land begins. The people who live there don't even try—they have come to terms with their environment, floating to and fro on the tide in their dugout canoes, or perched above it in their houses on stilts.

There are about twenty thousand of them and they call themselves the Asmat—which means "the people," "the human beings." Their nearest neighbors, some distance inland (and all outsiders), are known simply as Manowe—"the edible ones."

The Asmat are typically Papuan people, dark-skinned, woolly-haired, often heavily bearded, with wonderful arrow-shaped noses. Anthropologists classify them as Melanesoid—having social organizations built around men's houses and secret societies; marriage relationships involving the institutionalized exchange of goods; an accumulation of wealth in the form of pigs, stones, and shells; and a preoccupation with ancestral spirits and the skulls of the dead. All this is true, but it misses the point. What matters most is that the Asmat have developed an astonishingly complex and sophisticated social system, which not only produces some of the most expressive folk art to be found anywhere in the world, but also represents one of the most delicate and beautifully balanced ecological adaptations ever devised.

They eat each other.

There is no soil and there are no seasons on the Casuarina Coast. Over 5 meters of rain thunder down on the forest every year and run off into the shallow Arafura Sea through a series of streams and rivers connected by a network of capillary channels. The people hunt and fish along these waterways, taking an occasional crocodile or tree kangaroo, but the staple food and the center of their lives is the sago palm.

Metroxylon rumphii grows wild in the swampland. It is a tall and stately palm with a terminal crown of feathery leaves that don't hang down idly like those of a coconut palm, but reach alertly up into the air. It is one of those few amazing plants that seem to have been designed especially for our benefit. The leaves are tough and leathery and make durable containers or long-lasting shingles for hut roofs. Old ones make a handy tinder. The fibers in the leaves provide thread and a simple cloth for women's skirts. Branches scoop easily into trays and plates for serving food or, tied together by their own fiber, form the walls of the long houses. The spiky thorns on the outside of the trunk are used as needles and tools, but what really matters is the pith within, which can be pulped and leached to produce up to 50 kilograms of nourishing starchy sago.

Population pressure in Irian is low and nobody ever goes hungry, not just because the sago palm is quite common, but because every tree is recognized and protected by an elaborate pattern of land use. Every village, and every family in that village, has its traditional sago territories handed down from father to son, and even the payment of bride price is widely recognized as little more than a fee paid for the rights to use the wife's father's sago area under certain circumstances. The day after any wedding, the groom goes with his bride's family to their sago area so that he may make an immediate inspection tour of his potential new property.

The allocation, gathering, and preparation of sago are all hedged around with rituals, none of which are more important than the decapitation of the palm, which is dealt with at first as an enemy, or the subsequent "dressing" of the fallen trunk to give it the semblance of an impregnated mother who will soon be delivered of her life-giving fruits. This close identification between headhunting and childbirth, between old death and new life, is not accidental. The Asmat recognize that the resources of their environment, while bountiful, are not unlimited—and there is ready acceptance of the necessity for a succession of generations. "The man is dead," they say; "long live the child."

Headhunting, in essence, has little or nothing to do with war. It is the formal and ritual expression of a perceived need to keep things in balance. It is an Asmat admission of human responsibility for human destiny, a radical, but nevertheless highly realistic, solution to the problem of overpopulation. And whatever else you may feel about its taking place, you have to admit that it works. Or at least that it did until the missionaries arrived.

The first outsider to touch on the Casuarina Coast seems to have been Captain James Cook on his original voyage to the Pacific in 1770. He landed near the village of Pirimapun and, misunderstanding the intent of the greeting ceremony—which involves a number of canoes, each propelled by eight enormous men with bones in their noses, drumming on the wooden sides of the dugout, chanting, and hurling clouds of finely powdered lime into the air—gave an order to open fire. There is no record of the number of Asmat who were killed, but twenty of Cook's crew died in the ensuing battle.

Later contacts with Dutch administrators, and more recently with Japanese and Indonesian officials, have been only marginally more successful—reaching a peak of well-meaning misunderstanding with the arrival in 1958 of a small group of American Crozier missionaries. All of these foreigners have felt obliged to impose alien belief systems on the Asmat and have tried, without exception, to prohibit headhunting. All seem to have been incapable of understanding the subtle and pivotal position that headhunting plays in Asmat ecology—and all have, happily and wonderfully, failed so far to stop it entirely.

The Asmat continue to live and die largely by their own timetables.

Despite what happened to Captain Cook, the Asmat never kill haphazardly. When an accidental killing took place during my first visit twenty years ago, I discovered that there were rigorous social controls available to deal with almost any situation.

The whole Asmat area is divided up into pairs of villages that form traditional rivalries, each concentrating most of its headhunting activity on the other. Killings between them are frequent and take place as required, but this particular death was unintended and unscheduled. A man from Omadesep was hunting in the forest and, mistaking a movement in the undergrowth for a wild cassowary bird, speared a solitary fisherman from neighboring Atsj. The people were appalled. The situation between the villages was in equilibrium and nobody wanted to be involved in an untimely spiral of revenge. There is no place in the Asmat system for meaningless feuds, so they found another solution.

The killer was designated as "The Man Who Could Not Be Seen" and simply ceased to exist. But before he disappeared from social view, he was allowed to nominate a friend to act as go-between. This "Man Who *Could* Be Seen" arranged to visit Atsj and, in discussion with elders and the murdered man's family, agreed on the payment of appropriate compensation. The sum involved was considerable, including thousands of arrows, several bows and stone axheads, quantities of beads, and a number of steel knives and blades. It was far more than the killer could afford, but over the next days the whole village of Omadesep banded together to raise the fine on his behalf. And when it was ready, the dead man's family came to receive their due.

They arrived to find that "The Man Who Could Not Be Seen" was still invisible. He was walled up, along with his forfeit, in a hut with no doors or windows, and the visitors settled down in a circle around it. When all was ready, a hand suddenly thrust its way through the thatching holding a stone tool, the first of the agreed objects, and the dead man's wife came forward to receive it. But just before she was able to take the gift, it was pulled back inside as though the giver was of two minds. "I'm sorry I killed your husband," he was saying, in effect, "but don't get the idea that I am someone who can be pushed around. I am, after all, a man of Omadesep."

Eventually the stone exchanged hands and the family got up to go, but just as they were about to get back into their canoes, a great shout went up from the crowd. "There's another gift!" they cried, and back everyone came to go through the whole pantomime once again. Handing over the complete compensation took the entire day, with the last item changing hands just before sundown. The family from Atsj were actually in their canoes and had gone almost out of sight around a bend in the river when once more the cry went out. "Come back. We've found one more thing for you!" So back they came to receive the final gift—a beautiful ritual axhead carved entirely out of a single piece of quartz crystal. It was easily the most valuable object of all and one clearly in excess of the agreed sum, but the family recognized it for what it was, and took it and left. It was in fact an inspired device for finally healing the rift between the villages. The next day "The Man Who Could Not Be Seen" reentered society, became completely visible, and totally resolved the broken equilibrium by going across to Atsj to get the crystal back.

At the heart of this complex awareness of the importance of social stability lies the Asmat's notion of justice and their sense of right and wrong. It is always elegant, but seldom obvious. When a theft takes place, for instance, it is not a man's first concern to get the object back. The fact that he has had something stolen from him is embarrassing. It is a direct comment on his social standing, and his immediate reaction is to do everything possible to prove that he is a powerful man from whom things should not be so lightly taken. He starts by giving the thief (in a small community, everyone of course knows who was responsible) a gift of something even more valuable than the stolen goods. And this, naturally, is terribly embarrassing for the guilty man, who can only salvage his honor by giving the original victim something more expensive still . . . and so on. Theft is rare in Asmat.

And so is envy. The social code requires that you deal with anyone who admires or covets something that you own by giving it to them. Which is very effective, putting that person under an immediate obligation to you that can only be discharged by giving you something, or rendering you a service, of at least equal value.

There is even an arrangement for dealing with a man who covets his neighbor's wife. It is called *papisj.* When such a situation arises, the two men (both must be married) get together and agree to become papisj-partners, to engage in a ritual exchange of their wives. It is strictly forbidden for anyone to have an affair with the wife of someone who is not both in the village and in total agreement. And it is a relationship that cannot be undertaken casually, because papisj is very much more than wifeswapping. It involves the two men in lifelong obligations to come to each other's aid in all times of trouble, and puts a particularly heavy onus on the man who initiated the exchange—it is up to him to ensure that the husband of the woman he covets never comes to harm in battle.

Even the mechanics of the exchange itself are elegant. Once two men have agreed to papisj, they must individually approach their own wives and persuade them to cooperate. This is never easy. Even if the wives like the idea, they make sure that their approval depends on gifts and good behavior and always succeed in letting the delicate negotiations drag on for several days, during which they are showered with extra little attentions. Then, on the appointed day, the women exchange homes. Each goes to the papisj-partner's house and sits at his fireplace, preparing the evening sago and serving it to the entire family. After the meal, everyone else leaves and the two spend the night together. And early the next morning, each man decorates his partner's wife with red clay paint and feathers, and the woman in all her finery, bearing the gift of a new ax, returns triumphantly to her husband's home.

The Asmat have succeeded in turning what in other societies is a potentially divisive and destructive situation into a powerful force for social cohesion. Papisj reinforces the importance of reciprocity and extends both the contacts and the influence of those involved. Because the partners become blood kin, it creates new social bonds, cutting right across traditional and family ties or, in the case of distant relatives, strengthening relationships that might otherwise become obscured or forgotten. It is usually a private agreement between two men, but in times of group upheaval, it can serve an equally important public function. After a severe storm or a punishing war (or, on one occasion that I witnessed, when the community felt threatened by the arrival in their midst of a red-haired Australian with a wooden leg), everybody takes part. And in the early light, the village is vibrant with the passage of wonderfully decorated women, each proudly bearing a papisj present, returning to their own homes.

Kinship in Asmat is a complex matter—as it is in most tribal communities. There are proper names to cover every possible relationship, even those as distant as the female great-grandchildren of one's husband's great-grandparents (*ndoremus*), or as tangled as the cousin of one's second wife's younger brother (*ndoporoawis*). The Asmat believe that a child's body comes from its father, but its soul is taken directly from the spirit world. Soon after birth, every baby is taken to a "seer," who examines it very carefully, identifies and names the spirit involved, and protects the child against other spirits by rubbing the baby in her armpits, disguising its scent with her own.

From that moment on, the child is known by its "spirit name," unless sickness gives the family cause to invoke a more kindly spirit and a change of name. This continues until puberty, when children are thought to have acquired some degree of individuality and can be given a "human name," which derives usually from the father or grandfather and is taken very seriously. It is a name with power and responsibility, one known to the community, but never used lightly or in front of strangers. To protect the sanctity of the human name, everyone also has a nickname that refers to a title of honor or, more often, to an incident or characteristic that gives other people pleasure. Among my Asmat friends are Onesmanam—"The One Who Visits His Friends Only to Look at Their Wives"; Asamanampo—"He Who Constantly Looks Backward While Rowing"; and an old lady who, because of a single indiscretion in her youth, will always be known as Aotsjimai—"The Woman Who Gets Canoes Dirty with Her Feet Before the Lime Decoration Is Dry."

But more important than all of these is a special name that falls into the category of *owom*. An Asmat can have a spirit name, a human name, and a nickname, but still play no active role in any community unless he (or occasionally she) also has an owom name. This is the same as the human name of someone else from another village and one that can only be transferred following the death of that person. It is a "headhunting name."

When an Asmat comes of age and stands in need of initiation, preparations are made for a special ceremony. A giant canoe with extravagant carving is decorated and paddled off downstream by seven of the most influential elders, with the initiate sitting quietly in the prow. They travel through the swamps and out into the delta region, where the river mouth may be miles wide and fringed with thickets of mangrove. It is here, on the edge of Irian, that the shore is lined with bright-white broken shell and a thin line of whispering casuarinas. The canoe drifts with the tide across the mudflats and into the setting sun, keeping on going westward all through the night until it is 30 kilometers or more at sea, well out of sight of land.

At dawn the following day, the occupants turn and face back into the rising sun. When the first light touches the canoe, the initiate is curled into a fetal position on the floor with his eyes tightly closed. He is being born again. Steadily the elders paddle back into the new day and, as the coast comes into view, the initiate's eyelids flutter. His hands open, he pulls himself slowly up into a crawling posture, and when they enter the delta, he is sufficiently well coordinated to cling to the edge and look out at the view with wide childish eyes. Once the mangrove gives way to river forest, he is crouched on all fours, and as the canoe slips under the canopy into a channel that forms part of the village territory, he moves to sit upright. Fifteen years of development are concentrated into the next hour or two, so that by the time the boat turns the final bend in the stream and comes into full view of the village, the transition is complete and a grown man stands tall in the bow, acknowledging the applause of all the villagers, who gather along the muddy banks, drumming, dancing, and singing to welcome the arrival of a new warrior. And the name they use to greet him is his owom name, the name once worn with equal pride by the erstwhile owner of the skull he carries tucked underneath his arm. The head of a man who has had to die so that he might start to live.

The cycle is given substance by the fact that the initiate not only takes the name of the dead man, but becomes that man and takes on his relationships as well. The victim's family in another village use the name when talking to or telling about the initiate, and welcome him as a privileged mediator, who can move between the enemy villages, safe from any danger. They call him father, husband, or brother and extend to him the same care and concern they would display toward the man himself, had he lived. In truth, to them there is no ultimate difference. The dead man lives on, secure in the body of this stranger who has bored a small hole in the left temple of the skull, eaten his namesake's brains, and taken on his soul.

Parallel to this recognition of the survival of at least some part of a man killed in battle, there is a sense also of the responsibility for revenge. Of the existence of, and the need to do something about, a state of disequilibrium.

The memory of a murder may last a long time. It is kept alive by the erection of a distinctive totem pole called *bisj,* which looms over the village in constant and evocative reminder. The bisj is sculpted from the trunk of a mangrove tree that has steep buttress supports around its base. The crown of the tree is buried upside down several meters in the earth and carved into a series of figures standing upon each other's heads, decorated with hornbill and flying-fox motifs, with the uppermost embellished by a massive phallic framework cut from the rampant plank of the buttress itself. As long as this pole stands, retribution has yet to be exacted. But not just any killing will do. It has to be the right one at the appropriate time.

In 1957, in the final years of the Dutch colonial administration, two officials from Java made a visit to Asmat. They were minor bureaucrats, but in the fashion of the time traveled in grand style with an extensive company. Their tour included a call on the village of Otsjanep—one of the most traditional in Asmat—and they were greeted, as Captain Cook had been two centuries earlier, with an arrogant display. Like him, they misunderstood and opened fire, killing four of the most important men in the village—the elders with the human names of Osom, Foretbai, Akon, and Samat.

The Dutchmen and their party fled, without stopping to take the heads or learn the names of the murdered men, leaving Otsjanep in complete confusion. If the deaths had been the result of a raid by their traditional enemies from the neighboring village of Pirien, they would have known what to do, understood what was necessary to honor those who had died and restore essential balance. But these meaningless murders by unknown outsiders left them in a cultural vacuum, so they did the only thing possible. They erected four splendid bisj poles—"Lest We Forget"—and agreed to bide their time.

Four years later, Michael Rockefeller, great-grandson of the philanthropist John D. Rockefeller, came to Asmat to collect ethnic art for the Museum of Primitive Art in New York. His technique was simple and effective. He toured the villages with a party of dignitaries in motorized canoes, selecting the carvings he required, paying half the agreed price on the spot, and promising the rest on safe delivery of the goods to the mission headquarters at Agats. At the end of several months, an impressive collection had been gathered at Flamingo Bay. All the villages had kept their part of the bargain, except for Otsjanep—which had been mightily impressed by the size and importance of the Rockefeller retinue, and deeply moved that of all the sculpture in their village, the pieces the obviously powerful young leader, who was obliging enough to tell them his human name, most admired and wanted to buy were four carved tree trunks with the memories of Osom, Foretbai, Akon, and Samat.

Michael Rockefeller decided to return to Otsjanep with just a young anthropologist and two local boatmen for company. On the journey from Agats round to the next river mouth, their boat was swamped and they drifted out to sea. The boatmen swam to fetch help and when this failed to arrive, Rockefeller tied himself to two gasoline tanks for buoyancy and also set out for shore. The anthropologist was rescued the following day, still clinging to the boat, but despite massive sea and air searches by helicopters and landing craft of the United States Navy, the son of future Vice-President Nelson Rockefeller was never seen alive again.

In 1973, I happened to be in Otsjanep when the village was raided by a party from Pirien. A man from Pirien had died the day before in a skirmish between the two groups of hunters on a disputed creek. He was a villager with an owom name, a man without a wife and of no great consequence, but the killing was disturbing and had to be resolved. So the attack on Otsjanep came as no great surprise. It was expected and was anyway clearly announced by rude shouts from the undergrowth (something about the men of Otsjanep all having weak knees, like women), followed by a rain of arrows fired at random into the air.

I and all the women and children took cover in the houses and battle was joined on the mudflats alongside the river. It raged for about ten minutes with a huge noise as the braver souls in the front ranks hurled abuse and assorted missiles at each other. And then it stopped, almost as suddenly as it had started, as if by prearrangement and certainly by tacit agreement, as soon as a young warrior from Otsjanep was stabbed to death with a bone dagger. Honor was satisfied and the party from Pirien went happily home with his head.

That night there was a short funeral and a long feast. There had been several deeds of great valor in the battle and these had to be properly celebrated and compared with previous heroics. The bragging went on well into a drunken night, illustrated where appropriate with the trophies of other famous victories. I watched with polite interest as a succession of skulls taken from both friends and enemies (the friends can be recognized by the fact that they are still attached to their jaws) were produced for admiration. But toward the end of the evening I was electrified by the appearance of a skull with a difference. It was clearly European, male, about thirty years old at the time of death, with some expensive dental repairs—and a neat little hole drilled in the left temple.

"Eating people is wrong," we are told. And there are among us those who now do their best to pass this message on to the Asmat, along with an interesting selection of drug-resistant diseases—and various other benefits of civilization. There is a Catholic cathedral at Agats and a resident bishop who baptizes Asmat babies in a font modeled on a dugout canoe. The Indonesian government, which sent paratroops in to secure Irian in 1962, has passed and tries to enforce laws made in Djakarta forbidding the taking of heads and requiring the wearing of trousers. And when anyone is looking, the Asmat try to oblige—but it is difficult to convince someone who holds his grandmother's head in a place of honor in the house (keeping her informed of all the latest gossip, introducing her to the new children, and, when in need of spiritual guidance, even using her skull as a pillow) that this reverence is wrong and that the proper and Christian thing to do is to let the old lady rot in some swampy grave.

The Asmat never eat their relatives. There is no need to, ancestors are always with us anyway, but they deserve and get respect. And so do enemies. There is no point in hating someone from another village when you know that by killing him and eating part of him, you become him. The whole sequence of events that leads to headhunting and subsequent initiation is riddled with respect, both for enemies and for the environment. The Asmat have come to terms with both. Almost alone among the people of my acquaintance, they make ecological sense. They take their surroundings as they find them and, instead of trying to bend the world to suit their ends, modify their own behavior to make the very best of what is available. They have succeeded in developing a social system of astonishing intricacy and sophistication, one so sensitive to environmental pressures that it bends to accommodate every need and allows the Asmat to thrive in an area where we, with all our brute technology, must struggle to make a meager living.

The difference between us is best illustrated by something that happened during my last visit a few years ago. The Crozier mission at Agats is now quite substantial. There are stores and schools and sawmills on the one navigable river. A boardwalk connecting the landing stage and the various mission buildings allows the American priests to ply their trade without getting their feet wet and, to one side of the intrusive complex, beyond the last tin roof, an area of tidal swamp has been cleared to form an athletic field. Here, every Saturday afternoon, teams from the nearest villages meet each other in a noisy, mud-spattered, soggy, but nevertheless enthusiastic game of soccer.

This bush league has been going on now for some time and, despite the intense nature of the competition and the fierce struggles that ensue for possession of the ball, every single game for over seven years has ended in a draw. After a match that I watched, the referee—a newly arrived priest driven to the point of exasperation by his umpteenth game without a decision—remonstrated with the rival captains.

"Don't you see," he said, "that the object of the game is to try to *beat* the other team? Someone has to win!"

The two men looked at him with compassion, reconciled to the fact that he was young and had a lot to learn. They shook their heads firmly and said, "No, father. That's not the way of things. Not here in Asmat. If someone wins, then someone else has to lose—and that would never do."

The Asmat know best.

References

1. The Success of Failure

Eiseley, L. *The Immense Journey.* New York: Vintage, 1959.

Holloway, R.L. "Endocranial Capacities of the Early African Hominids . . ." *Journal of Human Evolution* 2:449, 1973.

Krantz, G.S. "Brain Size and Hunting Ability in Earliest Man." *Current Anthropology* 9:450, 1968.

Tobias, P.V. *The Brain in Hominid Evolution.* New York: Columbia University Press, 1971.

2. The Importance of Pattern

Cole, L.C. "Biological Clock in the Unicorn." *Science 125:874, 1957.*

Davies, P.C.W. The Accidental Universe. London: Cambridge University Press, 1982.

Dirac, P. *Directions in Physics.* London: Wiley, 1978.

Jung, C.J. *Synchronicity.* London: Routledge and Kegan Paul, 1972.

Kammerer, P. *Das Gesetz der Serie.* Stuttgart: DVA, 1919.

Koestler, A. *The Roots of Coincidence.* London: Hutchinson, 1972.

Levinson, H.C. *The Science of Chance.* London: Faber, 1952.

Stanford, R., and G. Thompson. "Unconscious Psi-meditated Instrumental Response and Its Relation to Conscious ESP Performance." In *Research in Parapsychology*, ed. W.G. Roll et al. Metuchen, N.J.: Scarecrow, 1973.

Weaver, W. *Lady Luck: The Theory of Probability.* New York: Dover, 1963.

Wheeler, J.A. *The Physicist's Conception of Nature.* Amsterdam: Reidel, 1978.

3. The Roots of Awareness

Backster, C. "Evidence of a Primary Perception in Plant Life." *International Journal of Parapsychology* 10:329, 1968.

Darwin, C. *Insectivorous Plants.* London: Murray, 1875.

Darwin, C. *The Power of Movement in Plants.* London: Murray, 1872.

Hashimoto, K. *Choshinrigaku Nyumon.* Tokyo: Kodansha, 1971.

Horowitz, K.A., et al. "Plant Primary Perception." *Science* 189:478, 1975.

Jones, C., and J. Wilson. "Sensitive Plants Trip on Negative Charge." *New Scientist* (29 July 1982):302.

Pickard, B.G. "Action Potential in Higher Plants. *Botanical Review* 39:172, 1973.

Simon, A. "A Theoretical Approach to the Classical Conditioning of Botanical Subjects." *Journal of Biological Psychology* 20:35, 1978.

Simons, P. "Anaesthetics Wake Up Dozy Plants." *New Scientist* (18 August 1983):472.

Thellier, M., et al. "Do Memory Processes Also Occur in Plants?" *Physiologia Plantarum* 56:281, 1982.

Tompkins, P., and C. Bird. *The Secret Life of Plants.* New York: Harper and Row, 1973.

4. The Nature of Crowds

Breder, C.M. "Fish Schools as Operational Structures." *Fishery Bulletin* 74:471, 1976.

Canetti, E. *Crowds and Power.* Harmondsworth, U.K.: Penguin, 1973.

Davis, J.M. "The Coordinated Aerobatics of Dunlin Flocks." *Nature* 309:344, 1984.

Heppner, F.H., and J.D. Haffner. "Communication in Bird Flocks." In *Biological and Clinical Effects of Low Frequency Magnetic and Electrical Fields,* ed. J.G. Llaurado et al. Springfield, Ill.: Thomas, 1974.

Lissaman, P.B.S., and C.A. Shollenberger. "Formation Flight of Birds." *Science* 168:1003, 1970.

Mackie, G.O. "Studies on *Physalia physalis*: Behaviour." *Discovery Reports* 30:369, 1960.

Pomeroy, H., and F.H. Heppner. "Laboratory Determination of Startle Reaction Time of the Starling." *Animal Behaviour* 25:720, 1977.

Potts, W.K. "The Chorus-Line Hypothesis..." *Nature* 309:344, 1984.

Selous, E. *Thought Transference (or What?) in Birds*. London: Constable, 1931.

Totton, A.K. "Studies on *Physalia physalis*: Natural History." *Discovery Reports* 30:301, 1960.

Woodcock, A.H. "Notes Concerning *Physalia* Behaviour at Sea." *Limnology and Oceanography* 16:551, 1971.

5. The Source of Concern

Morris, R.L. "Psi and Animal Behavior." *Journal of the American Society of Psychical Research* 64:242, 1970.

Osis, K. "A Test of the Occurrence of Psi Between Man and the Cat." *Journal of Parapsychology* 16:233, 1952.

Rhine, J.B., and S.P. Feather. "The Study of Cases of Psi-trailing in Animals." *Journal of Parapsychology* 26:1, 1962.

Schouten, S.A. "Psi in Mice." *Journal of Parapsychology* 36:261, 1972.

Schul, B. "The Psychic Power of Animals." New York: Fawcett, 1977.

Sheldrake, R. *A New Science of Life*. London: Blond and Briggs, 1981.

Stevenson, I. "Telepathic Impressions." *Proceedings of the American Society for Psychical Research* 29:1, 1970.

Targ, R., and K. Harary. *The Mind Race*. New York: Villard, 1984.

Wylder, J.E. *Psychic Pets*. London: Dent, 1980.

6. The Dreams of Dragons

Auffenberg, W. *The Behavioral Ecology of the Komodo Monitor*. Gainesville: University Presses of Florida, 1981.

Bright, M. "Meet Mokele-mbembe." *Wildlife* 2:596, 1984.

Heuvelmans, B. *On the Track of Unknown Animals*. London: Paladin, 1970.

Mackal, R.P. "Searching for Hidden Animals." New York: Doubleday, 1980.

Maclean, P. *A Triune Concept of the Brain and Behavior.* Toronto: University of Toronto Press, 1973.

7. The Arteries of the Earth

Boshier, A. "Mining Genesis." *Mining Survey* 64:21, 1969.
Eitel, E.J. *Feng-shui.* London: Synergetic, 1984.
Mooney, J. "The Ghost Dance Religion and the Sioux." *Annual Report of the Bureau of American Ethnology* 14:641, 1896.
Pennick, N. *The Ancient Science of Geomancy.* London: Thames and Hudson, 1979.
Rossbach, S. *Feng-Shui.* New York: Dutton, 1983.
Skinner, S. *The Living Earth Manual of Feng Shui.* London: Routledge and Kegan Paul, 1982.

8. The Currents of Life

Becker, R.O., and A.A. Marino. *Electromagnetism and Life.* Albany: State University of New York Press, 1982.
Becker, R.O., and J.A. Spadaro. "Electrical Stimulation of Partial Limb Regeneration in Mammals." *Bulletin of the New York Academy of Medicine* 48:627, 1972.
Burr, H.S. *Blueprint for Immortality.* London: Spearman, 1972.
Burr, H.S. "The Meaning of Bioelectric Potentials." *Yale Journal of Biological Medicine* 16:353, 1944.
Burr, H.S., and C.T. Lane. "Electrical Characteristics of Living Systems." *Yale Journal of Biological Medicine* 8:31, 1935.
Dumitrescu, I. *Electrographic Imaging in Medicine and Biology.* London: Spearman, 1983.
Krippner, S., and D. Rubin, eds. *The Kirlian Aura.* New York: Doubleday, 1974.
Mackenzie, D. "The Electricity That Shapes Our Ends." *New Scientist* (28 January 1982):217.
Moss, T. *The Body Electric.* London: Granada, 1981.
Perry, F.S., et al. "Environmental Power-Frequency Magnetic Fields and Suicide." *Health Physics* 41:267, 1981.

9. The Wonder of Water

Alexandersson, O. *Living Water.* Wellinborough, U.K.: Turnstone, 1982.

Grad, B. "Some Biological Effects of the Laying-on of Hands." *Journal of the American Society for Psychical Research* 59:92, 1965.

Leopold, L.B., and K.S. Davis. "Water." Netherlands: Time-Life International, 1968.

Magat, M. "Change of Properties of Water Around 40° C." *Journal of Physical Radiometrics* 6:108, 1936.

Piccardi, G. *The Chemical Basis of Medical Climatology.* Springfield, Ill.: Thomas, 1962.

Pople, J. "A Theory of the Structure of Water." *Proceedings of the Royal Society* A202:323, 1950.

Schwenk, T. *Sensitive Chaos.* New York: Schocken, 1976.

10. The Immersion of Man

Hardy, A. "Was Man More Aquatic in the Past?" *New Scientist* 7:642, 1960.

Irving, L. "Respiration in Diving Mammals." *Physiological Review* 19:112, 1939.

La Lumiere, L.P. "Evolution of human bipedalism." *Philosophic Transactions of the Royal Society of London* B292:103, 1981.

Morgan, E. *The Aquatic Ape.* London: Souvenir, 1982.

Morgan, E. *The Descent of Woman.* London: Souvenir, 1972.

Morris, D. *The Naked Ape.* London: Cape, 1967.

11. The Biology of Bias

Barsley, M. *The Left-handed Book.* London: Souvenir, 1966.

Bruner, J. *On Knowing: Essays for the Left Hand.* Cambridge, Mass.: Harvard University Press, 1962.

Burt, C. *The Factors of the Mind.* London: University of London Press, 1940.

Geschwind, N., and P. Behan. *Proceedings of the National Academy of Science* 79:5097, 1982.

Gould, S.J. *The Mismeasure of Man.* New York: Norton, 1982.

Grist, L. "Why Most People Are Right-handed." *New Scientist* (16 August 1984):22.

Ornstein, R.E. *The Psychology of Consciousness.* London: Cape, 1972.

12. The Essence of Balance

Eyde, D. "Cultural Correlates of Warfare." Doctoral dissertation submitted to Yale University in 1967 after field work at Amanamkai in Asmat.

Gerbrands, A.A., ed. *The Asmat: Journals of Michael Clark Rockefeller.* New York: Museum of Primitive Art, 1967.

Index